Food Preparation and Preservation

Food Preparation and Preservation

Abhinit Kumar Singh

RANDOM

RANDOM PUBLICATIONS

NEW DELHI (INDIA)

Food Preparation and Preservation

ISBN 978-93-5111-524-3

Published in 2015 in India by

RANDOM PUBLICATIONS

4376-A/4B, Gali Murari Lal, Ansari Road
New Delhi-110 002
Phone : +9111-43580356, 011-23289044, 011-43142548
e-mail: sales@randompublications.com,
info@randompublications.com, randomexports@gmail.com

Reprinted 2022

Type Setting by : Friends Media, Delhi-110089
Digitally Printed at : Replika Press Pvt. Ltd.

Preface

Food preparation is the transformation of raw ingredients into food, or of food into other forms. It typically takes clean, harvested crops or butchered animal products and uses these to produce attractive, marketable and often long shelf-life food products. Food preparation dates back to the prehistoric ages when crude processing incorporated slaughtering, fermenting, sun drying, preserving with salt, and various types of cooking. Modern food preparation methods improve the quality of life for people with allergies, diabetics, and other people who cannot consume some common food elements.

Food preservation is the process of treating and handling food to stop or slow down food spoilage, loss of quality, edibility or nutritional value and thus allow for longer food storage. Preservation usually involves preventing the growth of bacteria, fungi, and other micro-organisms, as well as retarding the oxidation of fats which cause rancidity. Food preservation can also include processes which inhibit visual deterioration, such as the enzymatic browning reaction in apples after they are cut, which can occur during food preparation. Maintaining or creating nutritional value, texture and flavour is an important aspect of food preservation.

This book presents a comprehensive review of technologies, procedures and innovations in food preparation and preservation, stressing topics vital to the food industry today and pinpointing the trends in future research and development. This text is a must-have for scientists and engineers involved in food manufacture, research and development in both industry and academia, as well as students of food-related topics at undergraduate and postgraduate levels.

Author

Contents

1

Introduction

Food is any substance consumed to provide nutritional support for the body. It is usually of plant or animal origin, and contains essential nutrients, such as carbohydrates, fats, proteins, vitamins, or minerals. The substance is ingested by an organism and assimilated by the organism's cells in an effort to produce energy, maintain life, or stimulate growth. Historically, people secured food through two methods: hunting and gathering, and agriculture. Today, most of the food energy consumed by the world population is supplied by the food industry.

Food safety and food security are monitored by agencies like the International Association for Food Protection, World Resources Institute, World Food Programme, Food and Agriculture Organization, and International Food Information Council. They address issues such as sustainability, biological diversity, climate change, nutritional economics, population growth, water supply, and access to food.

Food Sources

Most food has its origin in plants. Some food is obtained directly from plants; but even animals that are used as food sources are raised by feeding them food derived from plants. Cereal grain is a staple food that provides more food energy worldwide than any other type of crop. Maize, wheat, and rice - in all of their varieties - account for 87% of all grain production worldwide. Most of the grain that is produced worldwide is fed to livestock.

Some foods not from animal or plant sources include various edible fungi, especially mushrooms. Fungi and ambient bacteria are used in the preparation of fermented and pickled foods like leavened bread, alcoholic drinks, cheese, pickles, kombucha, and yogurt. Another example is blue-green algae such as Spirulina. Inorganic substances such as salt, baking soda and cream of tartar are used to preserve or chemically alter an ingredient.

Plants

Many plants or plant parts are eaten as food. There are around 2,000 plant species which are cultivated for food, and many have several distinct cultivars.

Seeds of plants are a good source of food for animals, including humans, because they contain the nutrients necessary for the plant's initial growth, including many healthful fats, such as Omega fats. In fact, the majority of food consumed by human beings are seed-based foods. Edible seeds include cereals (maize, wheat, rice, et cetera), legumes (beans, peas, lentils, et cetera), and nuts. Oilseeds are often pressed to produce rich oils - sunflower, flaxseed, rapeseed (including canola oil), sesame, et cetera.

Seeds are typically high in unsaturated fats and, in moderation, are considered a health food, although not all seeds are edible. Large seeds, such as those from a lemon, pose a choking hazard, while seeds from apples and cherries contain a poison (cyanide).

Fruits are the ripened ovaries of plants, including the seeds within. Many plants have evolved fruits that are attractive as a food source to animals, so that animals will eat the fruits and excrete the seeds some distance away. Fruits, therefore, make up a significant part of the diets of most cultures. Some botanical fruits, such as tomatoes, pumpkins, and eggplants, are eaten as vegetables.

Vegetables are a second type of plant matter that is commonly eaten as food. These include root vegetables (potatoes and carrots), bulbs (onion family), leaf vegetables (spinach and lettuce), stem vegetables (bamboo shoots and asparagus), and inflorescence vegetables (globe artichokes and broccoli and other vegetables such as cabbage or cauliflower).

Animals

Animals are used as food either directly or indirectly by the products they produce. Meat is an example of a direct product taken from an animal, which comes from muscle systems or from organs. Food products produced by animals include milk produced by mammary glands, which in many cultures is drunk or processed into dairy products (cheese, butter, et cetera). In addition, birds and other animals lay eggs, which are often eaten, and bees produce honey, a reduced nectar from flowers, which is a popular sweetener in many cultures. Some cultures consume blood, sometimes in the form of blood sausage, as a thickener for sauces, or in a cured, salted form for times of food scarcity, and others use blood in stews such as jugged hare.

Some cultures and people do not consume meat or animal food products for cultural, dietary, health, ethical, or ideological reasons. Vegetarians do not consume meat. Vegans do not consume any foods that are or contain ingredients from an animal source.

Food Production

Most food has always been obtained through agriculture. With increasing concern over both the methods and products of modern industrial agriculture, there has been a growing trend toward sustainable agricultural practices. This approach, partly fueled by consumer demand, encourages biodiversity, local self-reliance and organic farming methods. Major influences on food production include international organizations (e.g. the World Trade Organization and Common Agricultural Policy), national government policy (or law), and war.

In popular culture, the mass production of food, specifically meats such as chicken and beef, has come under fire from various documentaries, most recently Food, Inc, documenting the mass slaughter and poor treatment of animals, often for easier revenues from large corporations. Along with a current trend towards environmentalism, people in Western culture have had an increasing trend towards the use of herbal supplements, foods for a specific group of person (such as dieters, women, or athletes), functional foods (fortified foods, such as omega-3 eggs), and a more ethnically diverse diet.

Several organisations have begun calling for a new kind of agriculture in which agroecosystems provide food but also support vital ecosystem services so that soil fertility and biodiversity are maintained rather than compromised. According the International Water Management Institute and UNEP, well-managed agroecosystems not only provide food, fibre and animal products, they also provide services such as flood mitigation, groundwater recharge, erosion control and habitats for plants, birds fish and other animals.

Taste Perception

Animals, specifically humans, have five different types of tastes: sweet, sour, salty, bitter, and umami. As animals have evolved, the tastes that provide the most energy (sugar and fats) are the most pleasant to eat while others, such as bitter, are not enjoyable. Water, while important for survival, has no taste. Fats, on the other hand, especially saturated fats, are thicker and rich and are thus considered more enjoyable to eat.

Sweet

Generally regarded as the most pleasant taste, sweetness is almost always caused by a type of simple sugar such as glucose or fructose, or disaccharides such as sucrose, a molecule combining glucose and fructose. Complex carbohydrates are long chains and thus do not have the sweet taste. Artificial sweeteners such as sucralose are used to mimic the sugar molecule, creating the sensation of sweet, without the calories. Other types of sugar include raw sugar, which is known for its amber color, as it is unprocessed. As sugar is vital for energy and survival, the taste of sugar is pleasant.

The stevia plant contains a compound known as steviol which, when extracted, has 300 times the sweetness of sugar while having minimal impact on blood sugar.

Sour

Sourness is caused by the taste of acids, such as vinegar in alcoholic beverages. Sour foods include citrus, specifically lemons, limes, and to a lesser degree oranges. Sour is evolutionarily significant as it is a sign for a food that may have gone rancid due to bacteria. Many foods, however, are slightly acidic, and help stimulate the taste buds and enhance flavor.

Salty

Saltiness is the taste of alkali metal ions such as sodium and potassium. It is found in almost every food in low to moderate proportions to enhance flavor, although to eat pure salt is regarded as highly unpleasant. There are many different types of salt, with each having a different degree of saltiness, including sea salt, fleur de sel, kosher salt, mined salt, and grey salt. Other than enhancing flavor, its significance is that the body needs and maintains a delicate electrolyte balance, which is the kidney's function. Salt may be iodized, meaning iodine has been added to it, a necessary nutrient that promotes thyroid function. Some canned foods, notably soups or packaged broths, tend to be high in salt as a means of preserving the food longer. Historically speaking, salt has been used as a meat preservative as salt promotes water excretion, thus working as a preservative. Similarly, dried foods also promote food safety.

Bitter

Bitterness is a sensation often considered unpleasant characterized by having a sharp, pungent taste. Dark, unsweetened chocolate, caffeine, lemon rind, and some types of fruit are known to be bitter.

Umami

Umami, the Japanese word for delicious, is the least known in Western popular culture, but has a long tradition in Asian cuisine. Umami is the taste of glutamates, especially monosodium glutamate or MSG. It is characterized as savory, meaty, and rich in flavor. Salmon and mushrooms are foods high in umami. Meat and other animal byproducts are described as having this taste.

Cuisine

Many cultures have a recognizable cuisine, a specific set of cooking traditions using various spices or a combination of flavors unique to that culture, which evolves over time. Other differences include preferences (hot or cold, spicy, etc.) and practices, the study of which is known as gastronomy. Many cultures have diversified their foods by means of preparation, cooking methods, and manufacturing. This also includes a complex food trade which helps the cultures to economically survive by way of food, not just by consumption. Some popular types of ethnic foods include Italian, French, Japanese, Chinese, American, Cajun, Thai, and Indian cuisine. Various cultures throughout the world study the dietary analysis of food habits. While evolutionarily speaking, as opposed to culturally, humans are omnivores, religion and social constructs such as morality, activism, or environmentalism will often affect which foods they will consume. Food is eaten and typically enjoyed through the sense of taste, the perception of flavor from eating and drinking. Certain tastes are more enjoyable than others, for evolutionary purposes.

Presentation

Aesthetically pleasing and eye-appealing food presentations can encourage people to consume foods. A common saying is that people "eat with their eyes". Food presented in a clean and appetizing way will encourage a good flavor, even if unsatisfactory.

Contrast in texture

Texture plays a crucial role in the enjoyment of eating foods. Contrasts in textures, such as something crunchy in an otherwise smooth dish, may increase the appeal of eating it. Common examples include adding granola to yogurt, adding croutons to a salad or soup, and toasting bread to enhance its crunchiness for a smooth topping, such as jam or butter.

Contrast in taste

Another universal phenomenon regarding food is the appeal of contrast in taste and presentation. Opposite flavors, such as sweet and saltiness, tend to go well together, such as in kettle corn and with nuts.

Food Preparation

While many foods can be eaten raw, many also undergo some form of preparation for reasons of safety, palatability, texture, or flavor. At the simplest level this may involve washing, cutting, trimming, or adding other foods or ingredients, such as spices. It may also involve mixing, heating or cooling, pressure cooking, fermentation, or combination with other food. In a home, most food preparation takes place in a kitchen. Some preparation is done to enhance the taste or aesthetic appeal; other preparation may help to preserve the food; others may be involved in cultural identity. A meal is made up of food which is prepared to be eaten at a specific time and place.

Animal Preparation

The preparation of animal-based food usually involves slaughter, evisceration, hanging, portioning, and rendering. In developed countries, this is usually done outside the home in slaughterhouses, which are used to process animals en masse for meat production. Many countries regulate their slaughterhouses by law. For example, the United States has established the

Humane Slaughter Act of 1958, which requires that an animal be stunned before killing. This act, like those in many countries, exempts slaughter in accordance to religious law, such as kosher, shechita, and dhabiha halal. Strict interpretations of kashrut require the animal to be fully aware when its carotid artery is cut.

On the local level, a butcher may commonly break down larger animal meat into smaller manageable cuts, and pre-wrap them for commercial sale or wrap them to order in butcher paper. In addition, fish and seafood may be fabricated into smaller cuts by a fish monger. However fish butchery may be done on board a fishing vessel and quick-frozen for preservation of quality.

Cooking

The term "cooking" encompasses a vast range of methods, tools, and combinations of ingredients to improve the flavor or digestibility of food. Cooking technique, known as culinary art, generally requires the selection, measurement, and combining of ingredients in an ordered procedure in an effort to achieve the desired result. Constraints on success include the variability of ingredients, ambient conditions, tools, and the skill of the individual cook. The diversity of cooking worldwide is a reflection of the myriad nutritional, aesthetic, agricultural, economic, cultural, and religious considerations that affect it.

Cooking requires applying heat to a food which usually, though not always, chemically changes the molecules, thus changing its flavor, texture, appearance, and nutritional properties. Cooking certain proteins, such as egg whites, meats, and fish, denatures the protein, causing it to firm. There is archaeological evidence of roasted foodstuffs at Homo erectus campsites dating from 420,000 years ago. Boiling as a means of cooking requires a container, and has been practiced at least since the 10th millennium BC with the introduction of pottery.

Cooking equipment

There are many different types of equipment used for cooking.

Ovens are mostly hollow devices that get very hot (up to 500 °F) and are used for baking or roasting and offer a dry-heat cooking method. Different cuisines will use different types of ovens; for example, Indian culture uses a Tandoor oven, which is a cylindrical clay oven which operates at a single high temperature. Western kitchens use variable temperature convection ovens, conventional ovens, toaster ovens, or non-radiant heat ovens like the microwave oven. Classic Italian cuisine includes the use of a brick oven containing burning wood. Ovens may be wood-fired, coal-fired, gas, electric, or oil-fired.

Various types of cook-tops are used as well. They carry the same variations of fuel types as the ovens mentioned above. Cook-tops are used to heat vessels placed on top of the heat source, such as a sauté pan, sauce pot, frying pan, or pressure cooker. These pieces of equipment can use either a moist or dry cooking method and include methods such as steaming, simmering, boiling, and poaching for moist methods, while the dry methods include sautéing, pan frying, and deep-frying.

In addition, many cultures use grills for cooking. A grill operates with a radiant heat source from below, usually covered with a metal grid and sometimes a cover. An open pit barbecue in the American south is one example along with the American style outdoor grill fueled by wood, liquid propane, or charcoal along with soaked wood chips for smoking. A Mexican style of barbecue is called barbacoa, which involves the cooking of meats such as whole sheep over an open fire. In Argentina, an asado (Spanish for "grilled") is prepared on a grill held over an open pit or fire made upon the ground, on which a whole animal or smaller cuts are grilled.

Raw Food Preparation

Certain cultures highlight animal and vegetable foods in their raw state. Salads consisting of raw vegetables or fruits are common in many cuisines. Sashimi in Japanese cuisine consists of raw sliced fish or other meat, and sushi often incorporates raw fish or seafood. Steak tartare and salmon tartare are dishes made from diced or ground raw beef or salmon, mixed with various ingredients and served with baguettes, brioche, or frites. In Italy, carpaccio is a dish of very thinly sliced raw beef, drizzled with a vinaigrette made with olive oil. The health food movement known as raw foodism promotes a mostly vegan diet of raw fruits, vegetables, and grains prepared in various ways, including juicing, food dehydration, sprouting, and other methods of preparation that do not heat the food above 118 °F (47.8 °C). An example of a raw meat dish is ceviche, a Latin American dish made with raw meat that is "cooked" from the highly acidic citric juice from lemons and limes along with other aromatics such as garlic.

Restaurants

Restaurants employ trained chefs who prepare food, and trained waitstaff to serve the customers. The term restaurant is credited to the French from the 19th century, as it relates to the restorative nature of the bouillons that were once served in them. However, the concept pre-dates the naming of these establishments, as evidence suggests commercial food preparation may have existed during the age of the city of Pompeii, and urban sales of prepared foods may have existed in China during the Song Dynasty. The coffee shops or cafés of 17th century Europe may also be considered an early version of the restaurant. In 2005, the population of the United States spent $496 billion for out-of-home dining. Expenditures by type of out-of-home dining were as follows: 40% in full-service restaurants, 37.2% in limited service restaurants (fast food), 6.6% in schools or colleges, 5.4% in bars and vending machines, 4.7% in hotels and motels, 4.0% in recreational places, and 2.2% in others, which includes military bases.

Food Manufacturing

Packaged foods are manufactured outside the home for purchase. This can be as simple as a butcher preparing meat, or as complex as a modern international food industry. Early food processing techniques were limited by available food preservation, packaging, and transportation. This mainly involved salting, curing, curdling, drying, pickling, fermenting, and smoking. Food manufacturing arose during the industrial revolution in the 19th century. This development took advantage of new mass markets and emerging new technology, such as milling, preservation, packaging and labeling, and transportation. It brought the advantages of pre-prepared time-saving food to the bulk of ordinary people who did not employ domestic servants.

At the start of the 21st century, a two-tier structure has arisen, with a few international food processing giants controlling a wide range of well-known food brands. There also exists a wide array of small local or national food processing companies. Advanced technologies have also come to change food manufacture. Computer-based control systems, sophisticated processing and packaging methods, and logistics and distribution advances can enhance product quality, improve food safety, and reduce costs.

Commercial Trade of Food

The World Bank reported that the European Union was the top food importer in 2005, followed at a distance by the USA and Japan. Food is now traded and marketed on a global basis. The variety and availability of food is no longer restricted by the diversity of locally grown food or the limitations of the local growing season. Between 1961 and 1999, there was a 400% increase in worldwide food exports. Some countries are now economically dependent on food exports, which in some cases account for over 80% of all exports.

In 1994, over 100 countries became signatories to the Uruguay Round of the General Agreement on Tariffs and Trade in a dramatic increase in trade liberalization. This included an agreement to reduce subsidies paid to farmers, underpinned by the WTO enforcement of agricultural subsidy, tariffs, import quotas, and settlement of trade disputes that cannot be bilaterally resolved. Where trade barriers are raised on the disputed grounds of public health and safety, the WTO refer the dispute to the Codex Alimentarius Commission, which was founded in 1962 by the United Nations Food and Agriculture Organization and the World Health Organization. Trade liberalization has greatly affected world food trade.

Marketing and Retailing

Food marketing brings together the producer and the consumer. It is the chain of activities that brings food from "farm gate to plate". The marketing of even a single food product can be a complicated process involving many producers and companies. For example, fifty-six companies are involved in making one can of chicken noodle soup. These businesses include not only chicken and vegetable processors but also the companies that transport the ingredients and those who print labels and manufacture cans. The food marketing system is the largest direct and indirect non-government employer in the United States.

In the pre-modern era, the sale of surplus food took place once a week when farmers took their wares on market day into the local village marketplace. Here food was sold to grocers for sale in their local shops for purchase by local consumers. With the onset of industrialization and the development of the food processing industry, a wider range of food could be sold and distributed in distant locations. Typically early grocery shops would be counter-based shops, in which purchasers told the shop-keeper what they wanted, so that the shop-keeper could get it for them.

In the 20th century, supermarkets were born. Supermarkets brought with them a self service approach to shopping using shopping carts, and were able to offer quality food at lower cost through economies of scale and reduced staffing costs. In the latter part of the 20th century, this has been further revolutionized by the development of vast warehouse-sized, out-of-town supermarkets, selling a wide range of food from around the world.

Unlike food processors, food retailing is a two-tier market in which a small number of very large companies control a large proportion of supermarkets. The supermarket giants wield great purchasing power over farmers and processors, and strong influence over consumers. Nevertheless, less than 10% of consumer spending on food goes to farmers, with larger percentages going to advertising, transportation, and intermediate corporations.

Prices

It was reported on March 24, 2008, that consumers worldwide faced rising food prices. Reasons for this development include changes in the weather and dramatic changes in the global economy, including higher oil prices, lower food reserves, and growing consumer demand in China and India. In the long term, prices are expected to stabilize. Farmers will grow more grain for both fuel and food and eventually bring prices down. Already this is happening with wheat, with more crops to be planted in the United States, Canada, and Europe in 2009. However, the Food and Agriculture Organization projects that consumers still have to deal with more expensive food until at least 2018.

It is rare for the spikes to hit all major foods in most countries at once. Food prices rose 4% in the United States in 2007, the highest increase since 1990, and are expected to climb as much again in 2008. As of December 2007, 37 countries faced food crises, and 20 had imposed some sort of food-price controls. In China, the price of pork jumped 58% in 2007. In the 1980s and 1990s, farm subsidies and support programs allowed major grain exporting countries to hold large surpluses, which could be tapped during food shortages to keep prices down. However, new trade policies have made agricultural production much more responsive to market demands, putting global food reserves at their lowest since 1983.

Food prices are rising, wealthier Asian consumers are westernizing their diets, and farmers and nations of the third world are struggling to keep up the pace. The past five years have seen rapid growth in the contribution of Asian nations to the global fluid and powdered milk manufacturing industry, which in 2008 accounted for more than 30% of production, while China alone accounts for more than 10% of both production and consumption in the global fruit and vegetable processing and preserving industry. The trend is similarly evident in industries such as soft drink and bottled water manufacturing, as well as global cocoa, chocolate, and sugar confectionery manufacturing, forecast to grow by 5.7% and 10.0% respectively during 2008 in response to soaring demand in Chinese and Southeast Asian markets.

Rising food prices over recent years have been linked with social unrest around the world, including rioting in Bangladesh and Mexico, and the Arab Spring.

As Investment

Institutions such as hedge funds, pension funds and investment banks like Barclays Capital, Goldman Sachs and Morgan Stanley have been instrumental in pushing up prices in the last five years, with investment in food commodities rising from $65bn to $126bn between 2007 and 2012, contributing to 30-year highs. This has caused price fluctuations which are not strongly related to the actual supply of food, according to the United Nations. Financial institutions now make up 61% of all investment in wheat futures. According to Olivier De Schutter, the UN special rapporteur on food, there was a rush by institutions to enter the food market following George W Bush's Commodities Futures Modernization Act of 2000. De Schutter told the Independent in March 2012: "What we are seeing now is that these financial markets have developed massively with the arrival of these new financial investors, who are purely interested in the short-term monetary gain and are not really interested in the physical thing – they never actually buy the ton of wheat or maize; they only buy a promise to buy or to sell. The result of this financialisation of the commodities market is that the prices of the products respond increasingly to a purely speculative logic. This explains why in very short periods of time we see prices spiking or bubbles exploding, because prices are less and less determined by the real match between supply and demand." In 2011, 450 economists from around the world called on the G20 to regulate the commodities market more.

Some experts have said that speculation has merely aggravated other factors, such as climate change, competition with bio-fuels and overall rising demand. However, some such as Jayati Ghosh, professor of economics at Jawaharlal Nehru University in New Delhi, have pointed out that prices have increased irrespective of supply and demand issues: Ghosh points to world wheat prices, which doubled in the period from June to December 2010, despite there being no fall in global supply.

Famine and Hunger

Food deprivation leads to malnutrition and ultimately starvation. This is often connected with famine, which involves the absence of food in entire communities. This can have a devastating and widespread effect on human health and mortality. Rationing is sometimes used to distribute food in times of shortage, most notably during times of war.

Starvation is a significant international problem. Approximately 815 million people are undernourished, and over 16,000 children die per day from hunger-related causes. Food deprivation is regarded as a deficit need in Maslow's hierarchy of needs and is measured using famine scales.

Food Aid

Food aid can benefit people suffering from a shortage of food. It can be used to improve peoples' lives in the short term, so that a society can increase its standard of living to the point that food aid is no longer required. Conversely, badly managed food aid can create problems by disrupting local markets, depressing crop prices, and discouraging food production. Sometimes a cycle of food aid dependence can develop. Its provision, or threatened withdrawal, is sometimes used as a political tool to influence the policies of the destination country, a strategy known as food politics. Sometimes, food aid provisions will require certain types of food be purchased from certain sellers, and food aid can be misused to enhance the markets of donor countries. International efforts to distribute food to the neediest countries are often coordinated by the World Food Programme.

Food Safety Issues

Foodborne illness, commonly called "food poisoning", is caused by bacteria, toxins, viruses, parasites, and prions. Roughly 7 million people die of food poisoning each year, with about 10 times as many suffering from a non-fatal version. The two most common factors leading to cases of bacterial foodborne illness are cross-contamination of ready-to-eat food from other uncooked foods and improper temperature control. Less commonly, acute adverse reactions can also occur if chemical contamination of food occurs, for example from improper storage, or use of non-food grade soaps and disinfectants. Food can also be adulterated by a very wide range of articles during farming, manufacture, cooking, packaging, distribution, or sale. These foreign bodies can include pests or their droppings, hairs, cigarette butts, wood chips, and all manner of other contaminants. It is possible for certain types of food to become contaminated if stored or presented in an unsafe container, such as a ceramic pot with lead-based glaze.

Food poisoning has been recognized as a disease since as early as Hippocrates. The sale of rancid, contaminated, or adulterated food was commonplace until the introduction of hygiene, refrigeration, and vermin controls in the 19th century. Discovery of techniques for killing bacteria using heat, and other microbiological studies by scientists such as Louis Pasteur, contributed to the modern sanitation standards that are ubiquitous in developed nations today. This was further underpinned by the work of Justus von Liebig, which led to the development of modern food storage and food preservation methods. In more recent years, a greater understanding of the causes of food-borne illnesses has led to the development of more systematic approaches such as the Hazard Analysis and Critical Control Points (HACCP), which can identify and eliminate many risks.

Recommended measures for ensuring food safety include maintaining a clean preparation area with foods of different types kept separate, ensuring an adequate cooking temperature, and refrigerating foods promptly after cooking. Foods that spoil easily, such as meats, dairy, and seafood, must be prepared a certain way to avoid contaminating the people for whom they are prepared. As such, the general rule of thumb is that cold foods (such as dairy products) should be kept cold and hot foods (such as soup) should be kept hot until storage. Cold meats, such as chicken, that are to be cooked should not be placed at room temperature for thawing, at the risk of dangerous bacterial growth, such as Salmonella or E. coli.

Allergies

Some people have allergies or sensitivities to foods which are not problematic to most people. This occurs when a person's immune system mistakes a certain food protein for a harmful foreign agent and attacks it. About 2% of adults and 8% of children have a food allergy. The amount of the food substance required to provoke a reaction in a particularly susceptible individual can be quite small. In some instances, traces of food in the air, too minute to be perceived through smell, have been known to provoke lethal reactions in extremely sensitive individuals. Common food allergens are gluten, corn, shellfish (mollusks), peanuts, and soy. Allergens frequently produce symptoms such as diarrhea, rashes, bloating, vomiting, and regurgitation. The digestive complaints usually develop within half an hour of ingesting the allergen.

Rarely, food allergies can lead to a medical emergency, such as anaphylactic shock, hypotension, and loss of consciousness. An allergen associated with this type of reaction is peanut, although latex products can induce similar reactions. Initial treatment is with epinephrine (adrenaline), often carried by known patients in the form of an Epi-pen or Twinject.

Other Health Issues

Human diet was estimated to cause perhaps around 35% of cancers in a human epidemiological analysis by Richard Doll and Richard Peto in 1981. These cancer may be caused by carcinogens that are present in food naturally or as contaminants. Food contaminated with fungal growth may contain mycotoxins such as aflatoxins which may be found in contaminated corn and peanuts. Other carcinogens identified in food include heterocyclic amines generated in meat when cooked at high temperature, polyaromatic hydrocarbons in charred meat and smoked fish, and nitrosamines generated from nitrites used as food preservatives in cured meat such as bacon.

Anticarcinogens that may help prevent cancer can also be found in many food especially fruits and vegetable. Antioxidants are important groups of compounds that may help remove potentially harmful chemicals. It is however often difficult to identify the specific components in diet that serve to increase or decrease cancer risk since many food, such as beef steak and broccoli, contain low concentrations of both carcinogens and anticarcinogens.

Diet

Dietary habits are the habitual decisions a person or culture makes when choosing what foods to eat. Many cultures hold some food preferences and some food taboos. Dietary choices can also define cultures and play a role in religion. For example, only kosher foods are permitted by Judaism, halal foods by Islam, and in Hinduism beef is restricted. In addition, the dietary choices of different countries or regions have different characteristics. This is highly related to a culture's cuisine.

Diet Deficiencies

Dietary habits play a significant role in the health and mortality of all humans. Imbalances between the consumed fuels and expended energy results in either starvation or excessive reserves of adipose tissue, known as body fat. Poor intake of various vitamins and minerals can lead to diseases that can have far-reaching effects on health. For instance, 30% of the world's population either has, or is at risk for developing, iodine deficiency. It is estimated that at least 3 million children are blind due to vitamin A deficiency. Vitamin C deficiency results in scurvy. Calcium, Vitamin D, and phosphorus are inter-related; the consumption of each may affect the absorption of the others. Kwashiorkor and marasmus are childhood disorders caused by lack of dietary protein.

Moral, Ethical, and Health Conscious Diets

Many individuals limit what foods they eat for reasons of morality, or other habit. For instance, vegetarians choose to forgo food from animal sources to varying degrees. Others choose a healthier diet, avoiding sugars or animal fats and increasing consumption of dietary fiber and antioxidants. Obesity, a serious problem in the western world, leads to higher chances of developing heart disease, diabetes, and many other diseases. More recently, dietary habits have been influenced by the concerns that some people have about possible impacts on health or the environment from genetically modified food. Further concerns about the impact of industrial farming (grains) on animal welfare, human health, and the environment are also having an effect on contemporary human dietary habits. This has led to the emergence of a movement with a preference for organic and local food.

Nutrition and Dietary Problems

Between the extremes of optimal health and death from starvation or malnutrition, there is an array of disease states that can be caused or alleviated by changes in diet. Deficiencies, excesses, and imbalances in diet can produce negative impacts on health, which may lead to various health problems such as scurvy, obesity, or osteoporosis, diabetes, cardiovascular diseases as well as psychological and behavioral problems. The science of nutrition attempts to understand how and why specific dietary aspects influence health.

Nutrients in food are grouped into several categories. Macronutrients are fat, protein, and carbohydrates. Micronutrients are the minerals and vitamins. Additionally, food contains water and dietary fiber.

The body is designed by natural selection to enjoy sweet and fattening foods for evolutionary diets, ideal for hunters and gatherers. Thus, sweet and fattening foods in nature are typically rare and are very pleasurable to eat. In modern times, with advanced technology, enjoyable foods are easily available to consumers. Unfortunately, this promotes obesity in adults and children alike.

References

Katz, Solomon. (2003). *The Encyclopedia of Food and Culture*, Scribner.

Kripke, Gawain. (2005). *Food aid or hidden dumping*?. Oxfam International, March.

Marion Nestle: *Food Politics: How the Food Industry Influences Nutrition and Healt*h, University Presses of California.

Regmi, Anita. (2001).*Changing Structure of Global Food Consumption and Trade.* Market and Trade Economics Division, Economic Research Service, USDA, May 30.

2

Food and Nutrition

The science of nutrition concerns everything that the body does with food in order to live, function, grow, and heal. Foods that are eaten on a regular basis make up a diet. Although geographic location and family traditions play major parts in forming dietary habits, food choices vary from person to person. To a high degree, what people eat affects their health and their enjoyment of life. In turn, healthy habits and a happy outlook can improve the ways in which the human body makes use of food. With the help of the sun, plants make their food from chemicals in the earth, water, and air. Animals—human beings included—cannot make their own food but must eat plants, other animals, or both.In the early 1990s the long-standing, traditional basic four food groups, consisting of meat, dairy products, grains, and fruits and vegetables, were reworked into the more balanced and healthy food guide pyramid. This pyramid has as its base the grain group; on the second level are the fruit and vegetable groups; on the third level are the meat and dairy groups; and at the top is the fats, oils, and sweets group.

Food Pyramid

Grain Group

At the base of the food guide pyramid are breads, cereals, rice, pasta, and other foods made from grain. Human beings need more daily servings of these foods than any others because grain-based foods provide B vitamins, iron, carbohydrates, and some protein. Six or more daily servings are recommended. A serving, for example, is one slice of bread, one ounce (28 grams) of dry cereal, or one half cup of cooked pasta.

Fruit and Vegetable Group

The sources of most vitamins and minerals belong to these two groups. They also provide fiber, which contains no nutrients but aids in moving food through the digestive system. Three to five servings of vegetables and two to four servings of fruit should be eaten every day. One half cup of any fruit or vegetable counts as one serving.

Dairy and Meat Group

On this level are two groups of foods, such as milk, cheese, poultry, fish, and eggs, that come mostly from animals; notable exceptions are nuts and dried beans. These groups are quite high in protein, calcium, iron, phosphorus, the B vitamins, and zinc. Two to three servings from each group are recommended daily. Eight ounces (227 grams) of milk or 3 ounces (85 grams) of meat, for example, count as a serving.

Fats, Oils, and Sweets Group

The top of the pyramid includes foods that may add pleasure to eating but provide only calories and little nutritional benefit to one's diet. These foods include salad dressings, cream, butter, margarine, sugars, soft drinks, and candies. They should be eaten only sparingly. The American Heart Association recommends that no more than 30 percent of one's daily calories be derived from fat.

Nutrients

To function, the human body must have nutrients. The nutrients known to be essential for human beings are proteins, carbohydrates, fats and oils, minerals, vitamins, and water.

Proteins

Proteins are made of amino acids, small units necessary for growth and tissue repair. Protein is the body's most plentiful substance except for water and, possibly, fat. Animal foods such as meat, fish, poultry, milk, and eggs are rich in protein. Good plant sources of protein are beans, peas, nuts, bread, and cereals. Combining plant sources, such as peanut butter with whole-grain bread or rice with beans, provides excellent protein. So does combining plant and animal sources such as cereal and milk or macaroni and cheese.

Carbohydrates

Starches and sugars are carbohydrates, the main source of the body's energy. Carbohydrates account for about half of the calorie intake for most Americans and up to four fifths of the calories in diets of African and Asian peoples. Carbohydrate-rich foods are also the main sources of protein for most of the world. Rice, wheat, corn, and potatoes are common rich sources of carbohydrates. Sugars are not essential foods. They provide energy (calories) but no nutrients. For that reason sugar is called an "empty calorie" food. Occasional sweets are not harmful to a healthy, active person, but excessive sugar can lead to tooth decay when eaten between meals, especially in sticky snack foods that cling to the teeth.

Fats and Oils

Fats and oils (which are liquid fats) are a concentrated source of energy. Fats in the diet are necessary for good health. They make certain vitamins available for use in the body, they cushion vital organs, they make up part of all body cells, and they help to maintain body temperature. Fats also delay pangs of hunger because a food mixture containing fat remains longer in the stomach. Nutritionists distinguish between different types of dietary fats, or fats in food. Saturated fats usually are solid in form and of animal origin.

In many typical diets, meat fat is the main source. It is known that saturated fats can raise the level of cholesterol in the blood. Cholesterol is a natural waxy substance made by the body. It helps to form digestive juices and does other important work. It is present in the body no matter what is eaten. When the body cells cannot absorb any more cholesterol, any excess begins to accumulate in the walls of the blood vessels and gradually narrows them. This condition may lead to a heart attack or stroke.

Minerals

Minerals are neither animal nor vegetable; they are inorganic. Almost all foods contribute to a varied intake of essential minerals. Most minerals are easy to obtain in quantities required by the body. A major exception is iron for children under age 4 and adolescent girls and women in the childbearing years. These groups need more iron than a normal diet may provide. Iron helps to build red blood cells. It also helps the blood carry oxygen from the lungs to each body cell. Rich sources of iron are meat, especially liver; egg yolks; and dark green vegetables.

Everyone at every age needs calcium. This mineral builds bones and teeth, and it is necessary for blood clotting. The best sources are milk and hard cheese. Others are leafy greens, nuts, and small fishes—such as sardines—with bones that can be eaten. Phosphorus works with calcium to make strong bones and teeth. A diet that furnishes enough protein and calcium also provides enough phosphorus. Other important minerals are sodium, potassium, iodine, magnesium, zinc, and copper.

Vitamins

The discovery of vitamins began early in the 20th century. It is likely that some still are undiscovered. Eating a wide variety of foods ensures getting enough vitamins whether or not they are identified. All living things need vitamins for growth and health. The body either cannot manufacture them at all or cannot normally manufacture them in sufficient amounts, and so must absorb them from food. Each vitamin has specific roles to play. Many reactions in the body require several vitamins, and the lack or excess of any one can interfere with the function of another.

Fat-soluble vitamins: Four vitamins—A, D, E, and K—are known as the fat-soluble vitamins. They are digested and absorbed with the help of fats that are in the diet.

Vitamin A is needed for strong bones, good vision, and healthy skin. It is found both in dark green and yellow fruits and vegetables.

Vitamin D is essential for children because it helps calcium and phosphorus to form straight, strong bones and teeth. With direct sunlight on the skin, the body can manufacture its own vitamin D. Infants and young children often need a vitamin D supplement. Vitamin D is added routinely to most milk during processing.

Vitamin D is already known to be vital for bone health. However, a recent study suggested that it may be more useful than calcium in protecting older women from factures associated with osteoporosis. Researchers assessed relations between postmenopausal hip fracture risk and calcium, vitamin D, and milk consumption.

Women consuming 12.5 μg or more of vitamin D per day from food plus supplements had a 37% lower risk of hip fracture than did women consuming less than 3.5 μg/day. In addition, the risk of hip fracture was not influ-enced by the amount of calcium intake from milk or supplements in these women. The researchers concluded that "Because women commonly consume less than the recommended intake of vitamin D, supplement use or dark fish consumption may be prudent."

Vitamin E helps to protect vitamin A and red blood cells. It is found in a wide variety of foods, and almost everyone gets enough.

Vitamin K is one vitamin that is made within the human body—by bacteria that live in the intestinal tract. Small amounts are found as well in the green leaves of spinach, kale, cabbage, and cauliflower and also in pork liver.

Women with a low vitamin K intake may be at increased risk of bone fractures. The researchers studied self-reported dietary vitamin K intake and measured bone mineral density (BMD) of the hip and spine in more than 1,100 men and almost 1,500 women 29-86 years old. Dietary and supplemental intakes of vitamin K were assessed with the use of a food-frequency questionnaire. The researchers also accounted for such factors as age, body mass index, smoking status, alcohol use, physical activity score, and menopause status and current estrogen use among the women. Women with the lowest vitamin K intake had significantly lower mean BMD than those with the highest intake of vitamin K. These associations remained after controlling for variables and after stratification by age or supplement use. However, no significant association was found between dietary vitamin K intake and BMD in men.

Fat-soluble vitamins can be stored in the body for long periods. They are stored mostly in the fatty tissue and in the liver.

Water-soluble vitamins: The vitamin B group of several vitamins helps to maintain healthy skin and a well-functioning nervous system. B vitamins also help to convert carbohydrates into energy. Vitamin C, or ascorbic acid, is needed for building the connective tissue that holds body cells together. Vitamin C is essential for healthy teeth, gums, and blood vessels. It also helps the body to absorb iron. These water-soluble vitamins are not stored in the body for long. Good sources should be eaten every day.

Vitamin C is known to play an important role in the structure of the collagencomposed membrane. Supplementing pregnant women's diets with vitamin C would prevent levels from the nutrient in white blood cells, where it is stored, from dropping.

For the study, 52 women in their fifth month of pregnancy were given either a placebo or 100 mg of vitamin C each day for three months. As is typical during normal pregnancy, vitamin C levels in blood plasma decreased for all the women. However, the white blood cell concentration of the vitamin decreased only among women given the placebo.

Additionally, women who took vitamin C supplements experienced an increase in their white blood cell concentration of the vitamin. At delivery, less than 5% of the women who received vitamin C supplements experienced premature membrane rupture, in comparison to nearly 25% of women taking the placebo. The researchers concluded that vitamin C supplementation maintains stores of the nutrient in white blood cells and "may have value in preventing premature rupture."

Commenting on the study, one researcher said it "contributes another piece of evidence that vitamin C is important in the events that lead to premature rupture of the membranes." More studies with ample power to detect a difference are needed to confirm the role of vitamin C in preventing premature rupture,". Extra vitamin C should not just come in the form of pills. Pregnant women should also be sure to eat lots of vitamin C-containing fruits and vegetables, such as citrus fruits and broccoli.

Water

In order to live, every cell in the body must be bathed in water. Water takes an active part in many chemical reactions and is needed to carry other nutrients, to regulate body temperature, and to help eliminate wastes. Water makes up about 60 percent of an adult's body weight. Requirements for water are met in many ways. Most fruits are more than 90 percent water.

Calories

The body's most basic need is for energy. To get energy it needs food as a fuel and oxygen to burn it. The amount of energy foods can produce is measured in units called calories. A food calorie, or kilocalorie, is the amount of heat required to raise the temperature of 1 kilogram (2.2 pounds) of water 1 degree Celsius (1.8 degrees Fahrenheit).

The body changes the calories in food into energy, which is necessary for every act from blinking an eye to running a race. Energy is also used for the growing process, for rebuilding damaged cells, and for regulating body systems. The number of calories needed each day depends upon how much energy an individual's body uses. An active child can need more calories than an adult who works at a desk. The body needs more calories in cold weather to stay at an even temperature.

If a person takes in more food than required to meet the body's needs, the excess calories are converted to fat—a stored form of energy. That causes weight gain. Eating too little causes weight loss because the fat is used for energy. One pound (0.5 kilogram) of stored fat contains about 3,500 calories. If weight loss is advisable, the best way to lose is by eating less of high-calorie foods and getting more exercise.

For most people a safe limit for losing weight is 2 pounds a week. Strict dieting to lose weight should be attempted only under a physician's care. So-called quick and easy weight-loss diets are fads. Most are unbalanced, emphasising one type of food and excluding others. The more unbalanced the diet, the more dangerous it is. Knowledge of the body's nutritional needs helps to guard against promoters of fad diets and special "health" foods that have no advantages and that may even be harmful.

Functional Foods

In the past, foods were recognised as being nutritious based on their content of essential nutrients (fat, protein, carbohydrate, minerals and vitamins) that are needed for normal metabolic activity and bodily function. In recent years however, a growing body of scientific evidence has demonstrated that other food constituents (often referred to as phytochemicals and zoochemicals, when they come from plant and animal foods, respectively) may provide health benefits beyond basic nutrition and help prevent chronic diseases like cardiovascular disease, cancer and osteoporosis among others.

Examples of such disease-fighting food constituents include omega-3 fatty acids from fish, flaxseed, soybean, hempseed and canola oil, isoflavones from soya beans, carotenoids (like beta-carotene and lycopene) from carrots, tomatoes and other red/orange fruits and vegetables, sulforophane from broccoli, polyphenols from tea and wine and soluble fibre from oats and barley. Foods that may have physiological benefits and/or have the ability to reduce the risk of chronic disease beyond basic nutritional functions are called functional foods. Functional foods may be conventional foods that are consumed in their natural state (like fruits, vegetables, grains) novel foods that are modified or enriched with functional components to provide greater health benefits (like omega-3 eggs, phytosterol-enriched margarines). Other words that are often used to describe functional foods include health foods, medical foods, designer foods, foods for special dietary use and pharmafoods.

Some functional foods have been scientifically proven to provide health benefits beyond basic nutrition. For many others however, the supporting evidence is only preliminary in nature. Unfortunately, it can take several years and millions of dollars for the purported health benefits of a functional food to be substantiated by scientific evidence. Although some functional foods may provide health benefits, no single food is a "magic bullet" for maintaining health and preventing disease. Consuming a balanced diet consisting of a variety of foods is the best way to maintain optimal health.

Generally speaking, functional foods are considered safe to use. Many functional foods are conventional foods that have been consumed for hundreds of years without any significant safety concerns. Novel foods that have been modified in some way to convey a functional benefit must undergo a thorough safety assessment before they can be sold. Despite the apparent safety of functional foods however, some may have the potential to be hazardous, particularly when consumed in excess. Before making any major dietary changes, it is important to consult with a health care provider, especially if you are nursing or pregnant, have a chronic disease, are taking prescription medications or are about to undergo surgery. Once again, it must be emphasised that no single food is a "magic bullet" for maintaining health and preventing disease. A balanced diet consisting of a variety of foods is the safest and healthiest way to maintain optimal health.

Nutraceuticals

Nutraceuticals are products that are isolated or purified from foods and generally sold in medicinal forms not usually associated with food (tablets, capsules, drops) that may have physiological benefits and/or have the ability to reduce the risk of chronic disease beyond basic nutritional functions. Examples of nutraceuticals include fish oils, soy isoflavones and canola phytosterols. Since nutraceuticals are sold in medicinal form and not consumed as food, they are often classified as natural health products.

Natural health products are naturally occurring substances that are consumed for the purpose of diagnosing, treating or preventing illness or maintaining or promoting health. Natural health product substances include plant, algal, fungal, or animal materials or extracts thereof; vitamins; minerals; amino acids; essential fatty acids and probiotics. They are usually sold in dosage form (capsule, pill, tablets or liquid extracts). The natural health product definition includes traditional medicines and homeopathic preparations but excludes drugs and foods, among other substances. Since nutraceuticals are sold in medicinal form, they are sometimes classified as natural health products.

In the United States, the term 'dietary supplements' is used in place of natural health products. There is generally a need for further research to substantiate the purported health benefits of natural health products. For some natural health products, the scientific evidence is strong, but still inconclusive. For many others, the evidence is only preliminary in nature. Moreover, the chemical compositions and biological mechanisms of many natural health products are still poorly understood. Many natural health products vary significantly in chemical composition and therefore cannot be expected to produce consistent therapeutic effects.

B-complex Vitamins

B-complex vitamin products provide several or all of the B-vitamins, but unlike multivitamins they do not contain other nutrients. Often, the B-vitamins are included at relatively high potencies, and the product may have names like "B-complex 100" or "B-complex 50," hinting at their composition of 100 mg or 50 mg of each of several of the B-vitamins. Selection of a B-complex mixture rather than a multivitamin or multivitamin with minerals presupposes a need for or benefit from the B-vitamins but not from other nutrients. There is little to no evidence that such a supposition is correct.

A rationale frequently offered for use of a B-complex mixture is that these vitamins are necessary for energy metabolism. While it is generally true that several of B-vitamins are involved in energy production, some are not. Vitamin B-6 is also involved in metabolism of amino acids and many other substances. Folic acid and vitamin B-12 are involved in production of DNA and RNA, among other actions.

Effects on Health: Marginal or deficient intakes of each of the B-vitamins are not equally likely to occur. The RDAs of vitamins B-1, B-2, and B-6 are similar. The RDA for niacin is much more, nearly 20 mg/day, and that for B-12 is much less, only 2 mcg/day. There is no rationale for taking all of the B-vitamins in equal amounts. Obtain B-vitamins from a wide variety of foods and a multivitamin with minerals.

Beta-Carotene

Beta-carotene is one of a group of compounds called carotenoids that are part of the red, yellow, and orange pigments found in plant foods. Beta-carotene is a precursor of vitamin A. It is well known for its antioxidant properties. Beta-carotene is found in dark green leafy vegetables and yellow-orange fruits and vegetables.

Epidemiological studies during the past 25 years have indicated that diets high in carotenoid-rich fruits and vegetables, as well as high serum levels of beta-carotene, are associated with a reduced risk for several chronic diseases including heart disease, cataracts, and some cancers. However, three large-scale clinical trials concluded that beta-carotene provided no protection against lung cancer or cardiovascular disease. There is also evidence that the antioxidant activity of beta-carotene and other carotenoids at very high levels may shift into pro-oxidant activity inducing oxidative stress.

Dietary Recommendation: There is no dietary reference intake for beta-carotene and other carotenoids because the research is not sufficient enough to support any recommendation at this time. Strenuous exercise increases oxygen consumption with an increase in the production of free radicals. Betacarotene may protect tissues against oxidative damage. In most studies involving exercise induced lipid peroxidation, beta-carotene was used in combination with other antioxidants such as vitamin E and vitamin C, making it difficult to determine if beta-carotene or other substances are responsible for any positive effects.

Supplement Range:

a) Athletes who habitually consume low levels of antioxidants may be at greater risk for the harmful effects of free radicals.

b) Athletes should be encouraged to eat a balanced diet that includes frequent servings of carotenoid-rich fruits and vegetables.

c) While mega doses or long term supplementation of beta-carotene is not warranted, there is no evidence that small amounts of supplemental beta-carotene is harmful to any population, including athletes.

Multi-Vitamins with Minerals

Multivitamin products have several advantages as sources of supplemental nutrients-they provide a wide spectrum of nutrients in generally appropriate proportions and ratios, and they give reasonable confidence that the total intake of a specific nutrient is not inadequate because of improper variations in food intake. The term "multivitamins" is not specific, but many such products are formulated to contain a very wide spectrum of vitamins and essential minerals. Products that contain both multiple vitamins and multiple minerals provide the greatest assurance that the total intake of each vitamin or mineral meets established nutritional need for that nutrient.

Similarly, the words "high potency" and numerous similar terms do not have specific meanings. Federal regulations allow use of "high potency" on the label if two-thirds of the nutrients included are at 100 percent or more of the Daily Value (DV). Some products are labeled "high potency" for this reason, even though a few of the nutrients are present only 10 to 25 percent of the DV because of their bulk (such as calcium) or cost (such as biotin). Many products labeled "high potency," "mega," "ultra," "maximum strength" or the like contains some of the vitamins at 100 to 300 percent or so of the DV. Many multivitamin/mineral products contain substances that may be nutritionally useful or even essential, but for which no DV has been established. These are listed only with an amount per day, rather than a percent DV.

Although the typical diet provides adequate amounts of many nutrients and perhaps surplus amounts of some, marginal intakes of certain nutrients are common among some sport groups. Nutrient intakes less than the DV usually do not indicate clinical deficiency, but the intakes are marginal and improvement of the diet or supplementation are needed to remove any risk of deficiency. Furthermore, intakes at the recommended level of several nutrients may not be adequate for persons at risk of deficiency-those with increased needs or nutrient losses. Anyone considering taking vitamin and mineral supplements must decide on multivitamins/minerals, individual nutrient supplements, or both. For convenience and lower cost, multivitamin/mineral products offer a broad array of nutrients at modest doses, and without taking a large number of tablets or capsules.

For high intakes of specific vitamins or minerals, multi products generally are not a good approach. Nutrients such as calcium or vitamin E may be present in relatively low amounts that make it difficult to achieve the relatively high intakes that many persons want. Multivitamin/mineral products are an excellent way to assure maintenance intakes of vitamin A, vitamin D, most of the B-vitamins, and most minerals. The multivitamin should be checked to make sure that it contains at least 0.4 mg (400 mcg) of folic acid. The Food and Nutrition Board recommends 400 mcg of folic acid from fortified foods and/or supplements to reduce the risk of neural tube defects in women of childbearing age.

Helpful intakes of some vitamins and minerals can be best obtained through supplements that contain only that nutrient, or perhaps a mixture of a few. Vitamin E intakes of 100 to 800 IU are impossible to get from conventional foods and are in very few multivitamins. Vitamin C intakes of 100 to 1,000 mg can be obtained from conventional foods if the diet is selected for that purpose (lots of citrus fruits, etc.), but some people find doing this every day can be difficult. Calcium intakes of 1,000 to 1,300 mg can be obtained from consistently high intakes of milk and milk products (about 5 servings per day), but many people cannot routinely achieve this.

Some nutrients are formulated into simple "multi" mixtures such as calcium with vitamin D, calcium with a few other minerals, or antioxidant products that might contain vitamin E, selenium, vitamin D, and some carotenes or a few other substances. These simple formulas and single-nutrient supplements can be conveniently used to achieve high intakes of targeted nutrients.

Niacin

Niacin is required for energy metabolism (burning of fats, carbohydrates, amino acids, and alcohol), and for the synthesis and processing of many essential substances by the human body. The term "niacin" may be used specifically to describe nicotinic acid, but may also include "niacinamide" (nicotinamide). Niacin activity can be provided by its synthesis from the amino acid tryptophan. Sixty milligrams of tryptophan provides 1 mg of nicotinic acid, and thus typical dietary protein intakes provide 5 to 7 mg of "niacin equivalents." The RDA is 14 mg per day for women and 16 mg per day for men. Niacin is an appropriate component of multivitamins and Bcomplex supplements.

Nicotinic acid (but not nicotinamide) is an effective drug for lowering blood lipids and cholesterol when given at doses of several thousand milligrams. Nicotinic acid (but not nicotinamide) doses of approximately 50 mg or higher may cause a flushing reaction (vasodilation) that can be uncomfortable but is not hazardous. Gradual increases in intake may prevent this reaction.

Intakes that are effective as a cholesterol-lowering drug carry a recognisable risk of substantial side effects (gastrointestinal distress and liver effects). Therapeutic doses should be avoided unless under proper medical supervision and monitoring. Slow-release forms of nicotinic acid have greater potency as a drug and potential for adverse effects. Persons taking crystalline (not slow-release) niacin in doses of 50 mg or more should be aware of the flushing reaction.

The two coenzyme forms of niacin, NAD and NADP, participate in numerous metabolic activities and are central in energy production from carbohydrates, amino acids and fatty acids, as well as glycogen synthesis. A deficiency has the potential to affect both muscular and nervous function. Studies have shown that supplementation with niacin in high doses can suppress free fatty acid release through decreased lipolysis. This in turn, can result in a decreased availability of a major fuel source, causing the muscles to reply more on its glycogen stores which may impair prolonged exercise performance.

There is no evidence that the requirement for niacin is increased in exercise, and there is no need for niacin supplementation in exercise.

a) Consume foods rich in niacin (e.g., enriched cereals, pasta, meats, chicken and peanut butter).

b) Take a multivitamin that includes 100 percent of the DV for niacin.

c) Be aware of the potential side effects with higher doses.

Selenium: Selenium may help prevent cancer and possibly heart disease. In early 2002, USDA's Agricultural Research Service launched a study to research moderately high doses of selenium and their effects on human cardiovascular health, immune function, and reproductive health. More than 30 healthy men aged 18-45 are participating in the study. For one year, half the volunteers will take a daily capsule providing five and one-half times the recommended dietary allowance of selenium in the form of high-selenium yeast. The other half will receive a daily placebo-a capsule containing only yeast and no selenium.

Samples of blood, urine, semen, and other specimens will be provided for laboratory tests at regular intervals. Participants will also receive tests for cardiovascular function and several other health indicators. In addition, they will complete detailed records of their exercise, general health, and the foods they have eaten during specific 3-day periods. It is hoped that the project, already in its second phase, will provide new, detailed information about selenium. device level, linked to Profibus at the intermediate level and to Ethernet at the highest level.

Almost all modern instruments come with some network protocol built in, whether it is used or not. Most smart process instruments are HART-compatible. A food plant may need to integrate several protocols to accommodate the conveyorheavy and discrete device-heavy packaging area with the more process-oriented areas of a plant. In this respect, a food plant can require a more challenging control system than much larger process plants in which the nature of information is more uniform. The food industry has been relatively slow to adopt some of these technologies, according to those experts charged with introducing these technologies to food companies. In part, this is because there have not been many new food plants built, but it may also be because the industry is only now beginning to appreciate the advantages offered by the various networks. As older plants are replaced or upgraded, the control systems will become smarter and more sophisticated.

Safety of Natural Health Products

Natural health products whose chemical compositions and biological mechanisms are well understood can be standardised to contain specific amounts of active constituents, and can therefore be expected to produce consistent therapeutic effects. Despite a lack of scientific evidence to substantiate their purported benefits and inconsistency in chemical composition and quality, natural health products can still be sold under current laws and regulations, provided that they are safe and that they do not bear any unsubstantiated drug-like claims (claims of treating or preventing illness or disease) on their labels.

Instead of drug claims, manufacturers are often use indirect and often misleading "general well-being" statements that are not always substantiated by scientific evidence. Many natural health products have been used as traditional remedies for hundreds of years and are therefore presumed to be safe. Moreover, natural health products usually produce much fewer adverse side effects than conventional drugs. Despite their historical use and low incidence of adverse effects however, it should not be assumed that all natural health products are safe. There is often a lack of scientific data proving the safety of natural health products in the context in which most North Americans are using them today.

There have been a number of instances when natural health products have been removed from the market due to severe adverse events, some of which were attributed to interactions with conventional drugs. There have also seen instances when natural health products have been contaminated with pharmaceuticals and other foreign agents. In Canada, products bearing drug identification numbers (DIN) on their labels are manufactured in accordance with strict drug standards and are therefore less likely to be contaminated than products that do not bear DINs. Due to the potential hazards of using natural health products it is important to consult with a health care provider before beginning self-treatment, especially if you are nursing or pregnant, have a chronic disease, are taking prescription medications or are about to undergo surgery.

A growing body of scientific evidence suggests that functional foods and natural health products have the potential to contribute significantly to health maintenance and disease treatment and prevention. Although the safety and efficacy of some of these products have been substantiated with scientific evidence, for many others, the evidence is only preliminary in nature. Furthermore, there are a number of problems and inconsistencies with the developing North American functional foods/natural health product industry that must be resolved before the true benefits of functional foods and natural health products can be realised. As laws and regulations improve and research continues to separate fact from fiction, functional foods and natural health products are likely to play an increasingly important role in health maintenance and disease prevention in coming years.

Figs: An Example of Functional Food

Functional foods is still evolving but refers to foods that, by virtue of physiologically active components, provide benefits beyond basic nutrition and may prevent disease or promote health. Increasingly, manufacturers are including the names of nonnutritive components on their labels. The public is becoming more knowledgeable about these nonnutritive ingredients because of the large amount of publicity and articles in the mass media.

Today there is an ever increasing demand for foods that will increase our intake of food components that are perceived to be "protective against disease." In the future, genetic engineering will allow us to provide a variety of fruit or vegetables that is highly elevated in a certain compound. One example of such research is the work by the USDA and others in the development of a tomato high in lycopene, a carotenoid understood to be an antioxidant and anticancer substance.

In some varieties, the concentration of lycopene can be three times that of regular tomatoes. The genes responsible for the synthesis of carotenoids such as lycopene have been isolated. The genes have been inserted in an edible yeast and have produced lycopene and carotenoid at 0.1% dry weight of isolated cells. We are increasingly using foods as preventive agents to decrease our risk of deadly chronic diseases, including cancer and heart disease.

Figs (Ficus Carica) are one of the earliest fruits cultivated. Figs were dried and stored for later consumption. There was a fig tree in the Garden of Eden, and the fig is the most mentioned fruit in the Bible. Figs are mentioned in a Babylonian hymnbook dated about 2000 B.C. As early as 2900 B.C., in early Sumerian times, the medical use of figs was already being stressed. Legend has it that the Greek goddess Demeter first revealed to mortals the fruit of autumn, which they called the fig. It was the favorite fruit of Cleopatra, and she ended her life with an asp brought to her in a basket of figs.

Every inhabitant of Athens was a "philosykos," literally translated, "a friend of the fig." Early Olympic athletes were given figs as a training food and figs were given as laurels to the winners of the first Olympics as a "medal". The Wildlife Conservation Society of New York has recently determined that, in the animal and plant kingdom, the highcalcium content of wild figs makes them a "keystone" fruit, critical to the survival of other plants and animals.

Although considered a fruit, the fig is actually a flower inverted into itself. The seeds are the real fruit in figs. They are the only fruit to ripen fully and semidry on the tree. Native to areas from Asiatic Turkey to northern India, figs spread to all the countries around the Mediterranean. Today the United States, Turkey, Greece, and Spain are the primary producers of dried figs. The Spaniards brought figs to the Americas in the early 16th century. The missionary fathers brought them to California where they were planted from San Diego to Sonoma.

"Mission Fig" gets its name from this history. The Smyrna fig was imported from Turkey to California in 1882. It was given the hybrid name of Calimyrna in honor of the grower's home. California, specifically the San

Joaquin Valley, produces 100% of the dried figs grown in the United States. Fig trees reach fruit-bearing age in one to two years and have tremendous longevity.

Some trees planted at the beginning of California's commercial fig production are still bearing fruit almost 100 years later. There are four principal varieties of cultivated figs: the amber-colored Calimyrna; the dark, purplish Mission; the small but high-in-sugar Adriatic, and the uniquely white-totranslucent amber Kadota. The Calimyrna and Mission varieties represent almost two-thirds of the commercial production.

Dried figs are available in many forms: whole for the consumer and, in industrial products, as paste, concentrate, nuggets, powder, and diced forms. Potassium sorbate is added to dried figs to inhibit yeast fermentation and mold growth. Dried figs are processed to bring their moisture content up from 14-20% to as high as 30%. Compared to other dried fruits such as apricots or apples, color in the Calimryna variety may be stabilised by the addition of low levels of sulfur dioxide.

Nutrient Composition of Dried Figs

One serving of figs is 100 g, about ¼ cup, or about three Calimyrna or four to five Mission figs. Figs are fat free, sodium free and, like other plant foods, cholesterol free. Of the common fruits, figs have the highest overall content of minerals, and their calcium content per serving is second to oranges. Figs provide more fiber than all of the common fruits. A single serving contains 20% of the daily value of fiber. More than 28% of the fiber is the soluble variety. Soluble fiber has been shown to help control blood sugar and lower blood cholesterol by binding to it in the digestive tract.

The addition of a soluble-fiber supplement to the diet can aid in weight loss. Pasman studied obese women and found that average energy intake decreased significantly after fiber supplementation while hunger and satiety scores did not change. In a second study of subjects with low-energy intakes, hunger scores were significantly decreased after fiber supplementation. The authors concluded that, by facilitating compliance to a low-energy intake, fiber may be useful in the treatment of obesity.

Thus, figs and their soluble fiber may be of help with weight reduction, particularly considering the recent finding that over half of Americans, 97 million people, are overweight and 23% are grossly overweight, i.e., obese. The upward trend in adult obesity seen in the last 40 years has also been observed in children. Figs, as both a fruit and a snack, are an ideal addition to a child's diet and an adult's because they represent an excellent source of fiber and are naturally sweet.

Reducing the Risk of Cancer

It has been estimated that up to 70% of the incidences of cancers are related to diet. Major dietary hypotheses relate high fat consumption to breast and colorectal cancers; high alcohol intake to respiratory, gastrointestinal, breast, and liver cancers; and low fiber intake to colorectal cancer. One hundred twenty-eight of 156 epidemiological studies have shown an inverse correlation between the consumption of fruits and vegetables and the incidence of cancer.

There was a substantial reduction in risk for all cancers among those consuming a large amount of fruits and vegetables. Previously, it was thought that the antioxidant vitamins, vitamin C, E, and b carotene, were responsible for the beneficial effects of fruits and vegetables as shown by epidemiology studies across many cultures.

However, recent supplement studies have been disappointing with respect to that hypothesis. For instance, a Cancer Institute study on male smokers was halted because it was found that 5-8 years of vitamin E supplementation had no benefit. b carotene supplementation, in fact, was correlated with a significant increase in the incidence of lung cancer. No effect of supplementation with the antioxidants vitamin C, E, or Coenzyme Q10 was found with respect to oxidative DNA damage in smokers.

Out of almost 200 epidemiological studies, the relationship between cancer risk and fruit and vegetable intake is exceptionally strong and consistent. People who eat greater amounts of fruits and vegetables have about one-half the risk of getting cancer and have less mortality from cancer. It is the hypothesis of many scientists that the polyphenols in the fruits and vegetables are responsible for the reduced cancer-risk benefits they offer.

The polyphenols can act by several mechanisms to prevent cancer, such as carcinogen- blocking activities, antioxidant and free-radical scavenging, and antiproliferation/ antiprogression actions. In test tube studies, five classes of polyphenols were tested for their ability to inhibit the binding of the potent carcinogen benzopyrene to DNA in liver cells. These compounds were active after being fed to rats and subsequent exposure of the rats' liver cells to the carcinogen.

At least 22 studies have been conducted in which cancer has been experimentally induced (usually by means of a chemical carcinogen or irradiation) in animals fed vegetables or fruits. In most of these studies, it has been unequivocally found that the animals experienced fewer tumors, smaller tumors, fewer metastases, less DNA damage, higher levels of enzymes involved in the detoxification of carcinogens, or other outcomes indicative of lower cancer risk.

For an example, in one study, rats were given a diet with and without a vegetable-fruit mixture. The rats were exposed to a high-fat diet and a chemical carcinogen that induced colorectal cancer. Rats fed high-fat diets developed significantly more tumors than did rats fed low-fat diets, which indicates the underlying importance of a low-fat diet for lowering the risk of cancer. Also, a lower number of tumors were found in the rats fed vegetables and fruits along with the fat and a carcinogen.

Besides their ubiquitous polyphenols, figs have other compounds with anticancer activity, specifically benzaldehyde and the coumarins. Benzaldehyde has been used successfully to treat terminal human carcinomas. Of 57 evaluable patients, 19 responded completely and 10 responded with a greater than 50% regression in their cancers when treated with benzaldehyde. Treatment of squamous cell carcinomas with benzaldehyde induced the cancer cells to change into keratinized, normal squamous cells.

Coumarins are the major compounds isolated from the volatile extract of figs. In fact, the total dry-weight coumarinic content of figs is 0.5%. The furanocomarins identified in figs include angelicin, marmesin, psoralen, umbelliferone, and bergapten. Coumarins have also been used for the treatment of prostate cancer. Among cancers, the incidence of skin cancer is currently growing at the fastest rate.

The age standardised rates for melanoma have increased 60% compared to the rates in the 1970s, with the greatest proportional increase occurring among younger age groups. Psoralens, such as the fig compound angelicin, are currently being investigated for the treatment of skin cancer and have been recommended for clinical trials because they have low skin phototoxicity. These compounds produce free radicals and photoadducts of DNA that inhibit proliferation of the cancer cells.

References

Curley, S., and Mark. (1990). *The Natural Guide to Good Health,* Lafayette, Louisiana, Supreme Publishing.

Galdston, I. (1960). *Human Nutrition Historic and Scientific.* New York: International Universities Press.

Mahan, L.K. and Escott-Stump, S. eds. (2000). *Krause's Food, Nutrition, and Diet Therapy* (10th ed.). Philadelphia: W.B. Saunders Harcourt Brace.

Thiollet, J.-P. (2001). *Vitamines & minéraux.* Paris: Anagramme.

3

Food Processing

Food processing is the transformation of raw ingredients into food, or of food into other forms. Food processing typically takes clean, harvested crops or butchered animal products and uses these to produce attractive, marketable and often long shelf-life food products. Similar processes are used to produce animal feed.

Food processing dates back to the prehistoric ages when crude processing incorporated slaughtering, fermenting, sun drying, preserving with salt, and various types of cooking (such as roasting, smoking, steaming, and oven baking). Salt-preservation was especially common for foods that constituted warrior and sailors' diets, until the introduction of canning methods. Evidence for the existence of these methods can be found in the writings of the ancient Greek, Chaldean, Egyptian and Roman civilizations as well as archaeological evidence from Europe, North and South America and Asia. These tried and tested processing techniques remained essentially the same until the advent of the industrial revolution. Examples of ready-meals also exist from preindustrial revolution times such as the Cornish pasty and Haggis. During ancient times and today these are considered processing foods. Food processing has also helped create quick, nutritious meals to give to busy families.

Modern food processing technology in the 19th and 20th century was largely developed to serve military needs. In 1809 Nicolas Appert invented a vacuum bottling technique that would supply food for French troops, and this contributed to the development of tinning and then canning by Peter

Durand in 1810. Although initially expensive and somewhat hazardous due to the lead used in cans, canned goods would later become a staple around the world. Pasteurization, discovered by Louis Pasteur in 1862, was a significant advance in ensuring the micro-biological safety of food.

Food processing dates back to the prehistoric ages when crude processing incorporated slaughtering, fermenting, sun drying, preserving with salt, and various types of cooking (such as roasting, smoking, steaming, and oven baking). Salt-preservation was especially common for foods that constituted warrior and sailors' diets, until the introduction of canning methods. Evidence for the existence of these methods can be found in the writings of the ancient Greek, Chaldean, Egyptian and Roman civilizations as well as archaeological evidence from Europe, North and South America and Asia. These tried and tested processing techniques remained essentially the same until the advent of the industrial revolution. Examples of ready-meals also exist from preindustrial revolution times such as the Cornish pasty and Haggis. During ancient times and today these are considered processing foods. Food processing has also helped create quick, nutritious meals to give to busy families.

Modern food processing technology in the 19th and 20th century was largely developed to serve military needs. In 1809 Nicolas Appert invented a vacuum bottling technique that would supply food for French troops, and this contributed to the development of tinning and then canning by Peter Durand in 1810. Although initially expensive and somewhat hazardous due to the lead used in cans, canned goods would later become a staple around the world. Pasteurization, discovered by Louis Pasteur in 1862, was a significant advance in ensuring the micro-biological safety of food.

Types of Human Food

Humans are commonly believed to be omnivorous animals that can consume both plant and animal products. Evidence suggests that early Homo Sapiens employed hunting and gathering as their primary means of food collection. This involves combining stationary plant and fungal food sources with mobile animals which must be hunted and killed in order to be consumed.

Additionally, it is believed that humans have used fire to prepare food prior to eating since their divergence from Homo erectus, possibly even earlier. As at least ten thousand years ago, humans developed agriculture, which has continually improved and altered the way in which food is obtained. This lead to a variety of important historical consequences, such increased population, the development of cities, and the wider spread of infectious diseases. The types of food consumed, and the way in which they are prepared vary widely by time, location, and culture.

A selection of different foods, possibly complementary, eaten together comprises a meal. "A meal incorporates a number of contrasts, hot and cold, bland and spiced, liquid and semi-liquid, and various textures. It also incorporates cereals, vegetables, and animal proteins". Each person dining must use at least one untensil which will enter the mouth. A table must be used, a seating order put into effect, along with applied restrictions on movement or other types of behaviour. People often choose to eat together or with others.

Meals are considered to be for family, friends who are close, and guests who are to be honored. Often named and patterned, meals play a role in an important social occassion, such as the celebration of many key cultural and religious festivals. Food is not just used for nutrition or as a reward, but is a method of communication, or a code. "The meaning of a meal is found in a system of repeated analogies. Each meal carries something of the meaning of the other meals; each meal is a structured social event which structures others in its own image".

Each group, family or culture taking part in a meal activity uses symbolic structures within the mal to varying degrees. "If food is treated as a code, the messages it encodes will be found in the pattern of social relations being expressed". It can have symbolic meaning, and can be used to affect personal behaviour, or enforce protocol. "When he buys an item of food, consumes it, or serves it, modern man does not manipulate a simple object in a purely transitive fashion; this item of food sums up and trasnmits a situation; it constitutes an information; it signifies. That is to say that it is not just an indicator of a set of more or less conscious motivations, but that it is a real sign, perhaps the functional unit of a system of communication. By this means that not only the elements of display in food, such as foods involved in rites of hospitality, for all food serves as a sign among the members of a given society". Outside a social event, food can be used as means for nourishment for a single individual.

The number of meals in a day, their size, composition, when and how they are prepared and eaten varies greatly around the world. This diversity can be attributed to a number of local factors, including climate, ecology, economy, cultural traditions and industrialisation. In societies where the availability of food has risen above subsistence levels and beyond staple foods, food is also sold pre-prepared for immediate consumption in restaurants and other similar retail premises. In industrial societies, meals often contain a higher proportion of food of animal origin. Food eaten in smaller quantities between meals is regarded as snack food, which can be attributed to over-nutrition.

Acquisition of Food

Food is traditionally obtained through farming, ranching, and fishing, with hunting, foraging and other methods of subsistence locally important for some populations, but minor for others. In the modern era in developed nations, food supply is increasingly dependent upon agriculture, industrial farming, aquaculture and fish farming techniques. These techniques aim to maximise the amount of food produced while minimising the cost. The techniques include a reliance on mechanised tools, from the threshing machine and seed drill, to the tractor and combine. Developed tools have been combined with the use of pesticides to promote high crop yields and to combat insects or mammals which reduce yield.

More recently, there has been a growing trend towards more Sustainable agricultural practices. This approach—which is partly fuelled by consumer demand - encourages biodiversity, local self-reliance and Organic farming methods. Major influences on food production are international policy, e.g. the World Trade Organisation and Common Agricultural Policy, or law and war. Food for livestock is fodder and traditionally comprises hay or grain.

Food from Plant Sources

a) Seeds

- Cereals from grasses, including barley, maize, oats, rice, rye, and wheat
- Cereals from non-grasses, including buckwheat, amaranth, and quinoa
- Legumes, including beans, peas, and lentils
- Nuts, including coconuts, almonds, and pine nuts
- Oilseeds, including sesame, sunflower, and hemp

b) Vegetables
  - Root vegetables, including potatoes, cassava, and turnips
  - Leaf vegetables, including amaranth, spinach, and kale
  - Sea vegetables, including dulse, kombu, and dabberlocks
  - Stem vegetables, including bamboo shoots, nopales, and asparagus
  - inflorescence vegetables, including globe artichokes, broccoli, and daylilies
  - Fruit vegetables, including pumpkin, okra, and eggplant

c) Fruits

d) Herbs and spices

From Animals

- Dairy products, including milk
- Eggs, including roe and caviar
- Insects, including honey
- Meat, including beef, frogs' legs, gaegogi, goat, horse, kangaroo, lamb, mutton, pork, veal, rodents, human (i.e. cannibalism)
- Offal, including blood
- Poultry, including chicken, turkey, duck, goose, pigeon or dove, ostrich, emu, guinea fowl, pheasant, quail
- Seafood, including finfish such as salmon and tilapia, and shellfish such as mollusks and crustaceans
- Snails
- Game, this includes all animals hunted for food.

From neither animals or plants

- Salt
- Mushrooms, which are a type of fungi
- Seaweed, which is a protist
- Water, including mineral water and spring (water)

Food Prepration and Trade

Food being prepared in large quantities. Whilst some food can be eaten without preparation, many foods undergo some form of preparation for reasons of safety, palatability, or flavour. At the simplest level this may involve washing, cutting, trimming or adding other foods or ingredients, such as spices. It may also involve mixing, heating or cooling, pressure cooking, fermentation, or combination with other food. Most food preparation takes place in a kitchen. The preparation of animal-based food will usually involve slaughter, evisceration, hanging, portioning and rendering.

Early food processing techniques were limited by the available food preservation, packaging and transportation. Early food processing mainly involved salting, curing, curdling, drying, pickling and smoking. An early processed food product was cheese. During the industrialisation era in the 19th century, food manufacturing arose. This development took advantage of new mass markets and emerging new technology, such as milling, preservation, packaging and labelling and transportation. It brought the advantages of pre-prepared time saving food to the bulk of ordinary people who did not employ domestic servants.

Manufactured Food

— Drinks: beer, juice, soft drink, squash, wine.
— Bread is a staple food for many nations, being made of risen dough of wheat or other cereals.
— Cakes and cookies
— Cheese is a curdled milk product, of which many varieties exist.
— Dessert is a course, usually sweet, and generally served after the main course, e.g. Ice cream.
— French fries, Chips
— Functional food
— Jam and Jelly
— Pasta
— Pie
— Pizza
— Processed meats
— Sandwiches
— Salad
— Sauce

— Sausage
— Snack food: Confectionery, Potato chips, Chocolate, Hardtack, candy bars, junk food
— Soup
— Sugar
— Therapeutic foods
— Yogurt

Food Trade

Now food is traded on a global basis. The variety and availability of food is no longer restricted by the diversity of locally grown food or the limitations of the local growing season. In 1994 trade liberalisation began when over 100 countries became signatories to the Uruguay Round of the General Agreement on Tariffs and Trade which included an agreement to reduce subsidies paid to farmers. This is underpinned by the WTO enforcement of agricultural subsidy, tariffs, import quotas and settlement of trade disputes that cannot be bilaterally resolved. Where trade barriers are raised on the disputed grounds of public health and safety, the WTO refer the dispute to the Codex Alimentarius Commission, which was founded in 1962 by the United Nations Food and Agriculture Organisation and the World Health Organisation.

The sale of surplus food traditionally took place once a week when farmers took their wares on market day, into the local village market place. Here food was sold to grocers for sale in their local shops for purchase by local people. With the onset of industrialisation, and the development of the food processing industry, a wider range of food could be sold and distributed in distant locations. Typically early grocery shops would be counter-based shops, in which purchasers told the shop-keeper what they wanted, so that the shop-keeper could get it for them.

In the 20th century supermarkets were born. Supermarkets brought with them a self-service approach to shopping using shopping carts and were able to offer quality food at lower cost, through economies of scale and reduced staffing costs. This was sometimes known as 'pile it high' In the latter part of the 20th century, this has been further revolutionised by the development of vast warehouse sized out-of-town supermarkets, selling an extraordinarily wide range of food from around the world.

Unlike food processors, food retailing is a two-tier market in which a small number of very large companies control a large proportion of supermarkets. The supermarket giants wield great purchasing power over farmers and processors, and strong influence over consumers.

Food and Health

Food deprivation leads to malnutrition and ultimately starvation. This is often connected with famine, which involves the absence of food in entire communities. This can have a devastating and widespread effect on human health and mortality. In 2003 it was estimated that each year, 40 million people die of hunger worldwide. Rationing is sometimes used to fairly distribute food in times of shortage, most notably during times of war. Food deprivation is regarded as a deficit need in Maslow's hierarchy of needs and is measured using famine scales.

Food aid can benefit people suffering from a shortage of food. Conversely, badly managed food aid can cause problems by disrupting local markets, depressing crop prices and discouraging food production. Its provision, or threatened withdrawal, is sometimes used as a political tool to influence the politics of the destination country. International efforts to distribute food to the neediest countries are co-ordinated by the World Food Programme.

Foodborne illness, commonly called "food poisoning," is caused by bacteria, toxins, viruses and prions. Food poisoning has been recognised as a disease of man since as early as Hippocrates. Murder by food poisoning was used during the Roman Empire. In the Middle Ages all Royal Courts had food tasters. The sale of rancid, contaminated or adulterated food was commonplace until introduction of hygiene, refrigeration, and vermin controls in the 19th century.

Discovery of techniques for killing bacteria using heat and other microbiological studies by scientists such as Louis Pasteur contributed to the modern sanitation standards that we enjoy today. This was further underpinned by the work of Justus von Liebig whose work led to the development of modern food storage and food preservation methods. The two most common factors leading to cases of bacterial foodborne illness are cross-contamination of ready-to-eat food from other uncooked foods and improper temperature control.

Less commonly, acute adverse reactions can also occur if chemical contamination of food occurs, for example from improper storage, or use of non-food grade soaps and disinfectants. Food can also be adulterated by a very wide range of articles (known as 'foreign bodies') during farming, manufacture, cooking, packaging, distribution or sale. For example, pests, hairs, cigarette butts, wood chips, metal shards, plasters etc.

Understanding of the causes of food-borne-illnesses and more systematic techniques for their elimination has led to the development of commercial systems such as HACCP which can, if properly implemented, identify and eliminate many, but not all, possible risks. HACCP is well suited to identifying and controlling these potential food safety risks.

Some people have food allergies or sensitivities to foods which are otherwise wholesome to the majority of people. The amount of the food substance required to provoke a reaction in a susceptible individual can be minute. For instance, tiny amounts of food in the air, too minute to be smelled, have been known to provoke lethal reactions in sufficiently sensitive individuals. In theory, any food may provoke a reaction, however, this most commonly involves gluten, corn, shellfish (mollusks), peanuts, and soy. Most patients present with diarrhea after ingesting certain foodstuffs, skin symptoms (rashes), bloating, vomiting and regurgitation. The digestive complaints usually develop within half an hour of ingesting the allergen.

Rarely, food allergy can lead to anaphylactic shock: hypotension and loss of consciousness. This is a medical emergency. An allergen associated with this type of reaction is peanut, although latex products can induce similar reactions. Initial treatment is with epinephrine (adrenaline), often carried by known patients in the form of an Epi-pen.

Food allergy is thought to develop easier in patients with the atopic syndrome, a very common combination of diseases: allergic rhinitis and conjunctivitis, eczema and asthma. The syndrome has a strong inherited component; a family history of these diseases can be indicative of the atopic syndrome.

Dietary habits play a significant role in the health and mortality of all humans. Concerns about foodborne illness have long influenced diet. Traditionally humans have learned to avoid foods that induce acute illness. Some believe that this is the underlying rationale behind some traditional religious dietary requirements. Additionally, many people choose to forgo food from animal sources to varying degrees.

The nutrient content of diets in industrialised countries contain more animal fat, sugar, energy, alcohol and less dietary fiber, carbohydrates and antioxidants. Contemporary changes to work, family and exercise patterns, together with concerns about the effect of nutrition and overeating on human health and mortality are all having an effect on traditional eating habits. Physicians and alternative medicine practitioners may recommend changes to diet as part of their recommendations for treatment.

More recently, dietary habits have been influenced by the concerns that some people have about the chronic impact on health that arise through the consumption of genetically modified food. Further concerns about the impact of industrial farming on animal welfare, human health and the environment are also having an effect on contemporary human dietary habits. This has led to the emergence of a counterculture with a preference for organic and local food.

Health and Nutrition

Nutrition is the study of the relationship between diet and states of health and disease. Absent appropriate nutrition, functioning will be compromised and diseases can take hold, potentially resulting in death. Between the extremes of optimal health and death from starvation or malnutrition, there is an array of disease states that can be caused or alleviated by changes in diet. Both deficiencies, excesses and imbalances in the diet can have have negative impacts on health, resulting in disease states such as scurvy, obesity or osteoporosis. Also, excess ingestion of elements that have no apparent role in health may have toxic and potentially lethal effects depending on dose. The science of nutrition attempts to understand how and why specific aspects of diet have specific influences on health.

The human body comprises chemical compounds such as water, amino acids (proteins), fatty acids (lipids), nucleic acids (DNA/RNA), and carbohydrates (e.g. sugars). These compounds in turn consist of elements such as carbon, hydrogen, oxygen, nitrogen, and phosphorus, and may or may not contain minerals such as calcium, iron, and zinc. Minerals also ubiquitously occur in the form of salts and electrolytes. All of these chemical compounds and elements occur in various forms and combinations (e.g. hormones/vitamins, phospholipids, hydroxyapatite), both in the human body and in organisms (e.g. plants, animals) that humans eat.

The human body must necessarily comprise those elements that humans eat and absorb into the bloodstream. The digestive system, except in the unborn fetus, is the first step in helping to make the different chemical compounds and elements in food available for the trillions of cells of the body. In the digestive process of an average adult, about seven litres of liquid, known as digestive juices, exit the internal body and enter the lumen of the digestive tract. The digestive juices help break chemical bonds between ingested compounds as well as modulate the conformation and/or energetic state of the compounds/elements.

Yet many compounds/elements are absorbed into the bloodstream unchanged, though the digestive process helps to release them from the matrix of the foods where they occur. Any unabsorbed matter is eliminated in the feces. Only a minimal amount of digestive juice is eliminated this way; the intestines reabsorb most of it otherwise the body would rapidly dehydrate (hence the devastating effects of persistent diarrhea).

The specific types of compounds and elements that are absorbed by the body can be determined by comparing the waste to the food. The effect that the absorbed matter has on the body can be determined by finding the difference between the pre-ingestion state and the post-digestion state. The effect may only be discernible after an extended period of time in which all food and ingestion must be exactly regulated and all waste must be analysed. The number of variables involved in this type of experimentation is very high. This makes scientifically valid nutritional study very time-consuming and expensive, which accounts for why a proper science of human nutrition is rather new.

Eating a variety of fresh, whole (unprocessed) foods has proven hormonally and metabolically favourable compared to eating a monotonous diet based on processed foods. In particular, fresh, whole foods provide higher amounts and a more favourable balance of essential and vital nutrients per unit of energy, resulting in better management of cell growth, maintenance, and mitosis (cell division) as well as of appetite and energy balance. A generally more regular eating pattern (e.g. eating medium-sized meals every 3 to 4 hours) has also proven more hormonally and metabolically favourable than infrequent, haphazard food intake.

Foods and food products have been a major part of this revolution in consumer trends with increasingly personalised values of convenience, cost, packaging, and taste all emerging as distinguishing variables in the food marketplace. Now food is ready to take its next major step in value-personalising health. This step will not be easy and will require a renewed research and development effort to complement the knowledge base of traditional foods. Consumers can readily perceive and appreciate many of the inherent values of foods, such as taste, texture, and convenience, yet improved health is not immediately perceivable as being associated with particular foods.

Functional foods are those that provide benefits beyond basic nutrition, or that are functional at some physiological level. The identification of these foods varies, with recognition that they incorporate new technologies but may also include whole foods. Thus, functionality can emerge from the bioactive properties of ingredients in food formulations as well as from the food matrix of natural whole foods. From a market perspective, however, the description of functional foods varies substantially from that used in the scientific literature.

The global functional food market has been estimated at $47.6b, and growing, with trends to watch year by year. There appears to be strong agreement that the consumer wants better nutrition, and reports on trends indicate how consumers are responding to perceptions of benefits. If regulation demands a marriage between science and marketing, then a comparison between the market perceptions and science knowledge base is helpful. There is a good working knowledge of the functional properties of macronutrients, and a wide range of vitamins and minerals in foods. New bioactives are constantly being identified, but it will take some time to bring the science base of these components to the level of the former.

The functional properties of whole foods (and whole diets) are relatively under researched. There is variation in the understanding of physiological end-points and biomarkers of disease, albeit enough in some areas to support food innovation. These latter areas (eg lifestyle disease prevention) can also be seen to support a range of functional food categories. The science behind diet and heart disease is relatively strong, but may be less well developed in stress management and eye health.

In other areas, the perceived health benefit may not be linked directly to benefits to human physiology. For example foods that are 'free from' certain chemicals or production processes relate rather to a broader sense of health. Thus, from the marketing perspective, the consumer driven functional food market is not always referenced to nutritional science.

The coalescence of nutrition science and food technology can also be seen from changing perspectives in the past century. Beginning with basic sanitation through to preventing deficiencies (particularly of vitamins and minerals), and then excesses through an understanding of the dietchronic disease relationship, nutrition science is now focused on health and wellbeing, improving quality of life, preventing disease and reducing risk of disease. The rationale for functional foods is well recognised by nutrition practitioners, notably dietitians, where an aging population, increased healthcare costs, more aware and autonomous consumers, better scientific evidence and changes in food regulation lay the path for their effective use.

Ill health can be brought about by an imbalance of nutrients, producing either an excess or deficiency which in turn affects body functioning in a cumulative manner. Moreover, because most nutrients are, in some way or the other, involved in cell-to-cell signalling (e.g. as building block or part of a hormone or signalling 'cascades'), deficiency or excess of various nutrients affects hormonal function also indirectly. Thus, because they largely regulate the expression of genes, hormones represent a link between nutrition and how our genes are expressed, i.e. our phenotype.

The strength and nature of this link are continually under investigation, but observations especially in recent years have demonstrated a pivotal role for nutrition in hormonal activity and function and therefore in health. Mineral and/or vitamin (tocotrienol and tocopherol) deficiency or excess may yield symptoms of diminishing health such as goitre, scurvy, osteoporosis, weak immune system, disorders of cell metabolism, certain forms of cancer, symptoms of premature aging, and poor psychological health (including eating disorders).

As of 2005, twelve vitamins and about the same number of minerals are recognised as 'essential nutrients', meaning that they must be consumed and absorbed - or, in the case of vitamin D, alternatively synthesised via UVB radiation—to prevent deficiency symptoms and death. Certain vitamin-like substances found in foods, such as carnitine, have also been found essential to survival and health, but these are not strictly 'essential' to eat because the body can produce them from other compounds.

Moreover, thousands of different phytochemicals have recently been discovered in food (particularly in fresh vegetables), which have many discovered and yet to be discovered properties including antioxidant activity. Other essential nutrients include essential amino acids, choline and the essential fatty acids. In addition to sufficient intake, an appropriate balance of essential fatty acids—omega-3 and omega-6 fatty acids—has been discovered to be crucial for maintaining health. Both of these unique 'omega' long-chain unsaturated fatty acids are substrates for a class of eicosanoids known as prostaglandins.

Alpha-linolenic acid (LNA) serves as the building block for the less-inflammatory PGE3 series of prostaglandins, whereas linoleic acid (LA) (and specifically its product, arachidonic acid, AA) serves as the building block for either the PGE1 (anti-inflammatory) or the PGE2 (pro-inflammatory) series. (The omega-6 fatty acid LA is the building block for the omega-6 fatty acid AA, but AA can also be obtained directly in the diet). The conversions of AA into the respective prostaglandins (PGE1/PGE2) have importantly been discovered to be under hormonal control, as insulin up-regulates and glucagon down-regulates the function of the enzymes responsible for the conversions.

Because different types and amounts of food eaten/absorbed affect insulin, glucagon and other hormones to varying degrees, not only the amount of omega-3 versus omega-6 eaten but also the general composition of the diet is now known to determine health implications in relation to essential fatty acids, inflammation (e.g. immune function) and mitosis (i.e. cell division). Several lines of evidence indicate lifestyle-induced insulin malfunction, referred to as insulin resistance, as a decisive factor in many disease states. For example, insulin resistance is strongly linked to chronic inflammation, which in turn is strongly linked to a variety of adverse developments such as arterial microinjuries and clot formation (i.e. heart disease) and exaggerated cell division (i.e. cancer).

Hyperinsulinemia and insulin resistance (the so-called metabolic syndrome) are characterised by a combination of abdominal obesity, elevated blood sugar, elevated blood pressure, elevated blood triglycerides, and reduced HDL cholesterol. The state of overfatness/obesity clearly contributes to insulin resistance, which in turn can cause type 2 diabetes. Virtually all obese and most type 2 diabetic individuals have marked insulin resistance. Although the association between overfatness and insulin resistance is clear, the exact (likely multifarious) causes of insulin resistance remain less clear.

Importantly, it has been demonstrated that appropriate exercise, more regular food intake and reducing glycemic load all can reverse insulin resistance in overfat individuals. Overfatness can unfavourably alter hormonal and metabolic status via resistance to the hormone leptin, and a vicious cycle may occur in which insulin/leptin resistance and overfatness aggravate one another.

The vicious cycle is putatively fuelled by continuously high insulin/ leptin stimulation and fat storage, as a result of high intake of strongly insulin/leptin stimulating foods and energy. Both insulin and leptin normally function as satiety signals to the hypothalamus in the brain; however, insulin/ leptin resistance may reduce this signal and therefore allow continued overfeeding despite large bodyfat stores.

In addition, reduced leptin signalling to the brain may reduce leptin's normal effect to maintain an appropriately high metabolic rate. There is debate about how and to what extent different dietary factors—e.g. intake of processed carbohydrates, total protein, fat, and carbohydrate intake, intake of saturated and trans fatty acids, and low intake of vitamins/minerals—contribute to the development of insulin- and leptin resistance. In any case, analogous to the way modern man-made pollution may potentially overwhelm the environment's ability to maintain 'homeostasis', the recent explosive introduction of high Glycemic Index- and processed foods into the human diet may potentially overwhelm the body's ability to maintain homeostasis and health.

Antioxidants are another recent discovery. As cellular metabolism/ energy production requires oxygen, potentially damaging compounds known as radical oxygen species or free radicals may form. For normal cellular maintenance, growth, and division, these free radicals must be sufficiently neutralised by antioxidant compounds, such as certain vitamins (vitamin C, vitamin E, vitamin K and the aforementioned phytochemicals as well as other compounds, some of which the body itself produces.

Different antioxidants are now known to function in a cooperative network, e.g. vitamin C can reactivate free radical-containing glutathione or vitamin E by accepting the free radical itself, and so on. It is now also known that the human digestion system contains a population of a range of bacteria which are essential to digestion, and which are also affected by the food we eat. The role and significance of the intestinal bacterial flora is under investigation.

Since the Industrial Revolution some two hundred years ago, the food processing industry has invented many technologies that both help keep foods fresh longer and alter the fresh state of food as they appear in nature.

Cooling is the primary technology that can help maintain freshness, whereas many more technologies have been invented to allow foods to last longer without becoming spoiled. These latter technologies include pasteurisation, autoclavation, drying, salting, and separation of various components, and all appear to alter the original nutritional contents of food.

Pasteurisation and autoclavation (heating techniques) have no doubt improved the safety of many common foods, preventing epidemics of bacterial infection. But some of the (new) food processing technologies undoubtedly have downfalls as well. Modern separation techniques such as milling, centrifugation, and pressing have enabled upconcentration of particular components of food, yielding flour, oils, juices and so on, and even separate fatty acids, amino acids, vitamins, and minerals. Inevitably, such large scale upconcentration changes the nutritional content of food, saving certain nutrients while removing others.

Heating techniques may also reduce food's content of many heat-labile nutrients such as certain vitamins and phytochemicals, and possibly other yet to be discovered substances. Because of reduced nutritional value, processed foods are often 'enriched' or 'fortified' with some of the most critical nutrients (usually certain vitamins) that were lost during processing. Nonetheless, processed foods tend to have an inferior nutritional profile than do whole, fresh foods, regarding content of both sugar and high GI starches, potassium/sodium, vitamins, fibre, and of intact, unoxidised (essential) fatty acids.

In addition, processed foods often contain potentially harmful substances such as oxidised fats and trans fatty acids. A dramatic example of the effect of food processing on a population's health is the history of epidemics of beri-beri in people subsisting on polished rice. Removing the outer layer of rice by polishing it removes with it the essential vitamin thiamin, causing beri-beri. Another example is the development of scurvy among infants in the late 1800's in the United States. It turned out that the vast majority of sufferers were being fed milk that had been heat-treated to control bacterial disease. Pasteurisation was effective against bacteria, but it destroyed the vitamin C.

Lifestyle- and obesity-related diseases are becoming increasingly prevalent all around the world. There is little doubt that the increasingly widespread application of some modern food processing technologies has contributed to this development. The food processing industry is a major part of modern economy, and as such it is influential in political decisions. In any known profit-driven economy, health considerations are hardly a priority; effective production of cheap foods with a long shelf-life is more the trend.

In general, whole, fresh foods have a relatively short shelf-life and are less profitable to produce and sell than are more processed foods. Thus the consumer is left with the choice between more expensive but nutritionally superior whole, fresh foods, and cheap, usually nutritionally inferior processed foods. Because processed foods are often cheaper, more convenient (in both purchasing, storage, and preparation), and more available, the consumption of nutritionally inferior foods has been increasing throughout the world along with many nutrition-related health complications.

Most Governments provide guidance on good nutrition, and some also impose mandatory labelling requirements upon processed food manufacturers to assist consumers in complying with such guidance. Current dietary guidelines in the United States are presented in the concept of a food pyramid. There is no apparent consisteny in science-based nutritional recommendations between countries, indicating the role of politics as well as cultural bias in research emphasis and interpretation.

Nutritional Quality of Food

Food production, distribution, and consumption in the world lie at the intersection of diverse interests in nutrition, clinical medicine, public health, land use, economics, ecosystem health, labour concerns, immigration policies, justice, spirituality, and national security. Few issues are as fundamental, cross-cutting, and all-encompassing. Because of the wide reach of many aspects of food systems into personal and public life, the way that such a basic human need is met is the focus of attention of a large and diverse group of organisations and institutions.

Intersections with other perspectives are apparent and suggest opportunities as well as barriers to improving aspects of the food system in ways that benefit a variety of interests. Since diet is a leading contributor to morbidity and mortality, supporting healthy food systems in health care is an opportunity to address major causes of disease. Hospitals can play a leadership role in promoting healthy agriculture and healthy foods. This paper discusses a rationale and proposes some goals and expected benefits from moving hospitals and agriculture towards a healthier system.

At the outset it is important to acknowledge that food production and distribution systems have wide-ranging impacts on the quality of ecosystems, including their human communities. The nutritional attributes of food and food contaminants impact the consumer most directly. However, agricultural production can be and often is enormously destructive of ecosystems. Soil erosion, desertification, salinisation, water and air pollution, loss of habitat and biodiversity, and social and economic disruption also directly and indirectly negatively impact human and ecological health.

The biotic community is a dynamic, interconnected whole and that the health of any member of the community is dependent on the health of the whole. "Health" in this instance, as Aldo Leopold reminded us, is the capacity of the community to renew itself. One of the reasons that health care costs have been spiraling out of control is that we have ignored this fundamental principle of biology while focusing on only a small part of the system.

With some geographical variations food agriculture is characterised by declining numbers of small and mid-sized farms, increasing concentration of large industrial vertically-integrated agricultural systems producing for large commodity markets (e.g., corn, soybeans, sugar, pork, beef), and some increase in small producers for differentiated markets. Increasing concentration of people in large urban centers has led to re-design of food production systems and development of complex transportation systems to bring food to local markets. The entire system is increasingly dependent on fossil fuel consumption.

Moreover, many food products are redesigned to meet the needs of the system, with efficiency, durability, and marketability as drivers rather than with nutritional quality and socioeconomic community impacts in mind. As a result, food is both quantitatively and qualitatively different from that which was available decades ago, and the impacts of the food system on communities and the land are profound.

Despite significant advances in the nutritional sciences, large numbers of people do not eat a healthy diet. The composition of food products and the nature of the American diet are in large part a result of food production, distribution, and marketing methods, which are overwhelmingly based on products for large commodity markets.

Many preventable conditions, including various kinds of cancer, diabetes, coronary artery disease, obesity, low birth weight, and food-borne infectious diseases are directly linked to dietary factors, often in combination with genetic and other environmental contributors. These diseases are increasingly expensive to treat and are obvious targets for preventive measures.

Human Health Considerations

Despite claims about the high quality of food, many food analysts and public health professionals note with concern the increasing incidence of obesity, diabetes, food-borne illnesses, and other health conditions that are linked to nutritional factors and the food production and distribution system more generally. Confined animal feedlots provide an example of the relationship between food production systems and the nutritional quality of food. Beef cattle that are largely raised on corn in a feedlot, and routinely treated with antibiotics and hormones, reach marketable size more quickly than pastured animals. But the fat composition of the meat of the corn-fed animals contains a higher ratio of omega-6 to omega-3 fatty acids than grass-fed counterparts, in which healthier omega-3s are proportionately higher.

Antibiotic Use

Food agricultural systems that produce poultry, swine, beef, and farmed fish routinely use large amounts of antibiotics as growth promoters rather than as pharmaceutical agents to treat identified disease. In fact, more antibiotics are used in agricultural production than in clinical medicine. The routine use of antibiotics as growth promoters in animal husbandry contributes substantially to antibiotic resistance in bacteria that are human pathogens.

Pesticide Use

The widespread use of large amounts of pesticides in food production results in farm worker and farm community exposures, often at levels that exceed established "safety" limits. These exposures appear to increase risks of a variety of diseases, including some malignancies, neurodegenerative diseases, asthma, and birth defects.

Agricultural systems are major consumers of fossil-fuel derived fertilisers, pesticides, and growth-promoting pharmaceutical products. Confined animal feedlot operations (CAFOs) are point sources for run-off of growth promoters like arsenic, hormones, and antibiotics into local surface waters and in some cases into ground water. Studies of runoff from CAFOs show that biologically active hormones are present in surface waters at concentrations that are sufficient to alter fish reproduction and development. Organic arsenic, used as a growth promoter in swine and chicken production, not only contaminates the meat with arsenic at levels of concern but also is discharged into the environment when animal manure is spread onto the land.

Once in soil or sediments, organic arsenic is converted to its inorganic form, making it water-soluble and allowing it to seep into surface and groundwater ultimately used for drinking. Because of a dense concentration of animals in a relatively small space, CAFOs are also a source of noxious airborne emissions from manure lagoons that are sources of health concerns. Pesticide runoff and air emissions from agricultural operations contaminate waterways, rainwater, and air.

Over the past several decades, agricultural production and distribution has become increasingly concentrated in large, vertically-integrated systems supplying large commodity markets, though in some regions there has also been an increase in small producers of differentiated products for niche markets. These changes in agricultural production systems have had devastating impacts on smaller and mid-size farm families and farm communities in many areas. The implications for food security, public health, and the quality of ecosystems are profound. The commodity food system manufactures virtually all of its food products from 3 or 4 mass-produced commodities—led by corn and soybeans.

Small direct-marketing producers of differentiated products will never have the capacity to meet large market demands, and large commodity producers lack flexibility to address changing ecological circumstances and the need for nutritionally improved food. Experts in farming, food production, ecosystem management, and agricultural economics conclude that the trends are irreversible if significant action is not taken within the next 5-10 years. As small and mid-sized farms become less viable economically, development pressures often result in conversion of the land to other purposes.

Urban sprawl encroaches on farmland with resulting social and economic dislocation of families firmly rooted for generations. Entire regions become more dependent on importing food from greater and greater distances. Large industrial farming operations operate on exceedingly thin margins that are not economically sustainable over time and that depend on increasing size to survive over the short term. Animals raised for food in large industrial operations are usually confined in crowded conditions where they become sick and are injured. The ethical dimensions of animal husbandry practices are of significant and increasing concern.

Role for the Health Care System

A compelling rationale comes directly from the influence of dietary factors on health and disease risk. Hospitals routinely provide food for patients, staff, visitors, and the general public, affording a perfect opportunity to directly influence health and disease as well as to model dietary patterns for individuals and other institutions. One is reminded of the important role that hospitals played in the 1980s with the adoption of no-smoking policies as a demonstration of an important public health intervention.

In this case, by adopting food procurement policies that reflect an understanding of the importance of dietary factors and food production systems for human, ecological, and community health, health care institutions can address these concerns by showing leadership and providing markets. By adopting food procurement policies that support food production systems that are ecologically sound, economically viable, socially responsible, and morally feasible, health care systems also demonstrate an understanding of the inextricable links between human, public, and ecosystem health, or what might be collectively called "ecological health".

The health care system has a particular responsibility to address today's ecological realities because of the mission of the system, its opportunities, and the size of its ecological footprint. A new bioethics will see beneficence as doing good not only for the patient, but also for the entire community and the natural environment.

Adopt food procurement policies that provide nutritionally improved food for patients, staff, visitors, and the general public. Adopt food procurement policies that support food production systems that are ecologically sound, economically viable, socially responsible, and morally feasible. Adopt food procurement policies that reflect an ecological understanding of the dependence of human health on healthy ecosystems and that help to promote sustainable agricultural practices.

Expected benefits are the following::

— Health promotion and disease reduction
— Reduced use of non-therapeutic antibiotics in food production with decreased risk of antibiotic resistant organisms

— Reduced pesticide use; reduced pesticide exposures to farm workers, communities, and consumers
— Reduced ecological impacts of food production
— Improved social and economic conditions in food producing communities
— Improved hospital-community relations

Almost all food is processed in some way before it is eaten. Commercially, the main reasons to process food are to eliminate micro-organisms and to extend shelf life. Simply cooking or combining a food with other foodstuffs to create a recipe is also considered a form of food processing. Whatever the case, the nutrient value of any food is often altered by the processing. Some vitamins are more stable than others.

Water soluble vitamins (B-group and C) are more unstable than fat soluble vitamins (K, A, D and E) during food processing and storage. The most unstable vitamins include:

— Folate
— Thiamin
— Vitamin C.

More stable vitamins include:

— Niacin (vitamin B3)
— Vitamin K
— Vitamin D
— Biotin (vitamin B7)
— Pantothenic acid (vitamin B5).

There are a variety of things that happen during the growing, harvesting, storing and preparing of food that can affect its nutritional content.

— *Fertilisers*: Most plant crops are produced with the aid of fertilised soils. Fertiliser tends to reduce the vitamin C content of the crop, while increasing its protein value. It doesn't seem to make any difference to the plant's nutrient value whether the fertiliser is organic or not.
— *Milling*: Cereals, such as wheat, are ground to remove the fibrous husks. The husks contain most of the plant's dietary fibre, B-group vitamins, phytochemicals and some minerals. That's why products such as white bread are less nutritious than wholemeal varieties, even if they have

been artificially fortified with some of the nutrients that were lost after milling. It is impossible to add back everything that is taken out, especially the phytochemicals. The 'fibre' that is added back to some products is often in the form of resistant starch, which may not be as beneficial as the fibre removed.

— *Blanching*: Before a food is canned or frozen, it is usually heated very quickly with steam or water. The water soluble vitamins, including vitamin C and B-complex, are sensitive and easily destroyed by blanching.

— *Canning*: Food is heated inside the can to kill any dangerous micro-organisms and extend the food's shelf life. Some types of micro-organisms require severe heat treatment and this may affect the taste and texture of the food, making it less appealing. Preservatives are generally not needed or used in canned foods. Water soluble vitamins are particularly sensitive to high temperatures. Many people believe that canned foods are not as nutritious as their fresh counterparts; however, this is not always the case, as fresh food often deteriorates more rapidly than canned foods.

— *Freezing*: The nutrient value of a food is retained when it is frozen. Any nutrient losses are due to the processing prior to freezing and the cooking once the frozen food is thawed.

— *Pasteurisation*: This is the heating of milk to destroy micro-organisms. The nutrient value of milk is generally unaffected. In the case of pasteurised fruit juices, some losses of vitamin C can occur.

— *Dehydrating*: Drying out foods, such as fruits, can reduce the amount of vitamin C they retain but it can also concentrate other nutrients, particularly fibre in plant foods. Dehydrating food also makes food products more energy dense, which may contribute to weight gain. If a dehydrated food is reconstituted and cooked with water, further nutrients are leached out of the food and lost in the cooking water.

— *Preparation of vegetables*: Most vegetables are peeled or trimmed before cooking to remove the tough skin or outer leaves. However, the bulk of nutrients, such as vitamins, tend to lie close to the skin surface of most vegetables. Overzealous trimming can mean a huge reduction in a vegetable's nutrient value.

Some vitamins dissolve in water, so you lose your vitamins to the cooking water if you prefer to boil your vegetables. Alternative cooking methods—such as grilling, roasting, steaming, stir-frying or microwaving generally preserve a greater amount of vitamins and other nutrients.

It would be inaccurate to say that cooking food always lessens the nutrient value. Cooking can be advantageous in many ways, including:

— Making the food tastier.
— Breaking down parts of vegetables that would otherwise be indigestible.
— Destroying bacteria or other harmful micro-organisms.
— Making phytochemicals more available; for instance, phytochemicals are more available in cooked tomatoes than in raw tomatoes.

Preserving the nutrient value of vegetables: Some suggestions to retain the maximum nutrition in the foods you cook include:

— Store foods properly, such as keeping cold foods cold and sealing some foods in airtight containers.
— Keep vegetables in the crisper section of the refrigerator.
— Try washing or scrubbing vegetables rather than peeling them.
— Use the outer leaves of vegetables like cabbage or lettuce unless they are wilted or unpalatable.
— Microwave, steam, roast or grill vegetables rather than boiling them.
— If you boil your vegetables, save the nutrient-laden water for soup stock.
— Use fresh ingredients whenever possible.
— Cook foods quickly.

Labelling and Packaging and Nutrition

Nutrition is the process by which the foods we eat provide the nutrients we need to grow and stay healthy. Nutrients are naturally occurring chemical substances found in food. There are six categories of nutrients: proteins, lipids, carbohydrates, vitamins, minerals and water.

Proteins contain amino acids, sometimes referred to as the building blocks of protein. Dietary protein is supplied from plant and animal sources. Proteins are needed to build and repair body tissue and for the metabolic functions of our bodies. Lipids include fats and oils from plants and animals. Cholesterol is a fat found only in animal products. Lipids are of special interest because they are linked to the development of heart disease, the leading cause of death among the people.

The carbohydrates in our diet come from plant foods. Simple carbohydrates include the different forms of sugar, while complex carbohydrates include starches and dietary fiber. Vitamins are chemical compounds in our food that are needed in very small amounts (in milligrams and micrograms) to regulate the chemical reactions in our bodies. Minerals, also needed only in small amounts, have many different functions. Some minerals assist in the body?chemical reactions and others help form body structures.

Fifty to sixty percent of our body weight consists of water. It is the substance in which the metabolic reactions occur. We need about two quarts (2 liters) of water every day. Protein, fats and carbohydrates in food provide the energy, or kilocalories (kcals), our bodies need to function. Each gram of protein and carbohydrate has 4 kilocalories; each gram of fat has 9. You might have noticed that we use the metric system - grams, milligrams and micrograms—to measure the amounts of nutrients in foods.

Food labelling: Food labels for most of the food products sold in the markets must have the product name, the manufacturer name and address, the amount of the product in the package and the product ingredients. The ingredients are listed in descending order, based on their weight. Under the current laws, fresh fruits, vegetables and meat are exempt from these labelling requirements. In 1973 the Food and Drug Administration (FDA) established "nutrition labelling" or guidelines for labelling the nutrient and caloric content of food products.

Nutrition labelling is mandatory only for those foods that have nutrients added or make a nutritional claim. Manufacturers are encouraged, but not required, to provide nutrition labelling of other food products. Food labelling provides basic information about the ingredients in, and the nutritional value of, food products so that consumers can make informed choices in the market place.

Packaging: Food packaging development started with humankind?earliest beginnings. Early forms of packaging ranged from gourds to sea shells to animal skin. Later came pottery, cloth and wooden containers. These packages were created to facilitate transportation and trade. Utilising modern technology, today' society has created an overwhelming number of new packages containing a multitude of food products.

A modern food package has many functions, its main purpose being to physically protect the product during transport. The package also acts as a barrier against potential spoilage agents, which vary with the food product. For example, milk is sensitive to light; therefore, a package that provides a light barrier is necessary. The milk carton is ideal for that. Other foods like potato chips are sensitive to air because the oxygen in the air causes rancidity.

The bags containing potato chips are made of materials with oxygen-barrier properties. Practically all foods should be protected from filth, microorganisms, moisture and objectionable odors. Aside from protecting the food, the package serves as a vehicle through which the manufacturer can communicate with the consumer. Nutritional information ingredients and often recipes are found on a food label. The package is also utilised as a marketing tool designed to attract your attention at the store. This makes printability an important property of a package. The food industry utilises four basic packaging materials: metal, plant matter (paper and wood), glass and plastic.

A number of basic packaging materials are often combined to give a suitable package. The fruit drink box is an example where plastic, paper and metal are combined in a laminate to give an ideal package. This concept can be easily seen in your peanut butter jar. The main package containing the food (primary package) is made of glass (or plastic), the lid is made of metal lined with plastic, and the label is made of paper.

Each basic packaging material has advantages and disadvantages. Metal is strong and a good overall barrier, but heavy and prone to corrosion. Paper is economical and has good printing properties; however, it is not strong and it absorbs water. Glass is transparent, which allows the consumer to see the product, but breakable. Plastics are versatile but often expensive. Therefore, combining the basic materials works well in most cases. So, for a product like milk, which is an essential food for children and young adults and therefore cannot be very expensive, paper makes a good economical material. It also provides a good printing surface.

However, since paper absorbs water, it will gain moisture from the milk, get weaker and fail, thereby exposing the milk to spoilage factors. It may even break and waste the product. When a thin layer of a plastic called polyethylene is utilised to line the inside of the milk carton, it serves as a barrier to moisture and makes an economical, functional package. After making a food product and placing it in the appropriate package, a number of these individual packages must be placed in a large container to facilitate shipment. These larger containers are called secondary packages.

The paperboard box is a very common secondary package. Plastics also can serve as secondary packages. The milk case in which a number of milk cartons are delivered to the supermarket is a good example. Recycling is a sound approach. However, the problem often lies in feasibility of collection, separation and purification of the consumer disposed food packages. This mode of recycling is called post-consumer recycling. While it offers a logistic challenge, recycling is gaining in popularity, and the packaging industry is cooperating in that effort.

Aluminum cans are the most recycled container at this time. Plastic recycling is increasing, yet most plastic is recycled during manufacturing of the containers; not as post-consumer recycling. For example, trimmings from plastic bottles are reground and reprocessed into new ones. The plastics industry is helping to facilitate consumer recycling by identifying the type of plastic from which the container is made. A number from 1 to 7 is placed within the recycling logo on the container bottom.

For example, 1 refers to PET (Polyethylene Terephthalate), the plastic used for the large 2 liter soft drink bottles. Plastics have the advantage of being light. This helps to conserve fuel during transport and also reduces the amount of package waste. There are many interesting packaging concepts being explored by the industry to keep up with the changing life style of the consumer and new technologies. Many professionals are involved in designing and manufacturing the modern package. Today's package is designed with the consumer's safety and convenience in mind. Examples are microwaveable popcorn packages, squeezable ketchup bottles and the tamper-proof milk bottle cap.

A nutritionist must have a solid background in the physical, biological, behavioural and social sciences, along with an in-depth understanding of food composition. In addition, a nutritionist must be equipped with good management and communication skills in order to apply basic nutrition principles to practical aspects of dietetics and wellness.

Benefits of Food Processing

Benefits of food processing include toxin removal, preservation, easing marketing and distribution tasks, and increasing food consistency. In addition, it increases yearrly availability of many foods, enables transportation of delicate perishable foods across long distances and makes many kinds of foods safe to eat by de-activating spoilage and pathogenic micro-organisms. Modern supermarkets would not exist without modern food processing techniques, long voyages would not be possible and military campaigns would be significantly more difficult and costly to execute.

Processed foods are usually less susceptible to early spoilage than fresh foods and are better suited for long distance transportation from the source to the consumer. When they were first introduced, some processed foods helped to alleviate food shortages and improved the overall nutrition of populations as it made many new foods available to the masses.

Processing can also reduce the incidence of food borne disease. Fresh materials, such as fresh produce and raw meats, are more likely to harbour pathogenic micro-organisms (e.g. Salmonella) capable of causing serious illnesses.

The extremely varied modern diet is only truly possible on a wide scale because of food processing. Transportation of more exotic foods, as well as the elimination of much hard labour gives the modern eater easy access to a wide variety of food unimaginable to their ancestors.

The act of processing can often improve the taste of food significantly.

Mass production of food is much cheaper overall than individual production of meals from raw ingredients. Therefore, a large profit potential exists for the manufacturers and suppliers of processed food products. Individuals may see a benefit in convenience, but rarely see any direct financial cost benefit in using processed food as compared to home preparation.

Processed food freed people from the large amount of time involved in preparing and cooking "natural" unprocessed foods. The increase in free time allows people much more choice in life style than previously allowed. In many families the adults are working away from home and therefore there is little time for the preparation of food based on fresh ingredients. The food industry offers products that fulfill many different needs: From peeled potatoes that only have to be boiled at home to fully prepared ready meals that can be heated up in the microwave oven within a few minutes.

Modern food processing also improves the quality of life for people with allergies, diabetics, and other people who cannot consume some common food elements. Food processing can also add extra nutrients such as vitamins.

Drawbacks of Food Processing

Any processing of food can have slight effects on its nutritional density. Vitamin C, for example, is destroyed by heat and therefore canned fruits have a lower content of vitamin C than fresh ones. The USDA conducted a study in 2004, creating a nutrient retention table for several foods. A cursory glance of the table indicates that, in the majority of foods, processing reduces nutrients by a minimal amount. On average any given nutrient may be reduced by as little as 5%-20%.

Another safety concern in food processing is the use of food additives. The health risks of any additives will vary greatly from person to person; for example sugar as an additive would be detrimental to those with diabetes. In the European Union, only food additives (e.g., sweeteners, preservatives, stabilizers) that have been approved as safe for human consumption by the European Food Safety Authority (EFSA) are allowed, at specified levels, for use in food products. Approved additives receive an E number (E for Europe), which at the same time simplifies communication about food additives in the list of ingredients across the different languages of the EU.

Food processing is typically a mechanical process that utilizes large mixing, grinding, chopping and emulsifying equipment in the production process. These processes inherently introduce a number of contamination risks. As a mixing bowl or grinder is used over time the food contact parts will tend to fail and fracture. This type of failure will introduce in to the product stream small to large metal contaminates. Further processing of these metal fragments will result in downstream equipment failure and the risk of ingestion by the consumer. Food manufacturers utilize industrial metal detectors to detect and reject automatically any metal fragment. Large food processors will utilize many metal detectors within the processing stream to both ensure reduced damage to processing machinery as well risk to the consumer. The first industrial level metal detector pioneered by Goring Kerr was introduced back in 1947 for Mars Incorporated.

Performance Parameters for Food Processing

When designing processes for the food industry the following performance parameters may be taken into account:

— Hygiene, e.g. measured by number of micro-organisms per ml of finished product
— Energy efficiency measured e.g. by "ton of steam per ton of sugar produced"
— Minimization of waste, measured e.g. by "percentage of peeling loss during the peeling of potatoes'
— Labour used, measured e.g. by "number of working hours per ton of finished product"
— Minimization of cleaning stops measured e.g. by "number of hours between cleaning stops"
— De-agglomerating batter mixes in food processing

Problems often occur during preparation of batter mixes because flour and other powdered ingredients tend to form lumps or agglomerates as they are being mixed during production. A conventional mixer/agitator cannot break down these agglomerates, resulting in a lumpy batter. If lumpy batter is used to enrobe products, it causes an unsatisfactory appearance with misshapen or oversize products that do not fit properly into packaging. This can force production to a standstill. Furthermore batter mix is generally recirculated from an enrobing system back to a holding vessel; lumps then have a tendency to build up, reducing the flow of material and raising potential sanitation issues.

Using a high shear in-line mixer in place of a conventional agitator or mixer can quickly solve problems of agglomeration with dry ingredients. A single pass through a self-pumping, in-line mixer adds high shear to batter, which de-agglomerates the mix, resulting in a homogeneous, smooth batter. With a consistent, smooth batter, finished product appearance is improved; the effectiveness and hygiene of the recirculation system is increased; and a better yield of raw materials is achieved. By increasing overall product quality, the amount of raw materials needed is decreased, thereby lowering manufacturing costs.

High shear in-line mixers process food to be made faster and cheaper while increasing consistency of the finished food. Powder and liquid mixing systems are capable of rapidly incorporating large quantities of powders at high concentrations – agglomerate free and fully hydrated. Advances in technology have made processing equipment easy to clean, leading to a much safer processed food.

References

Davidson, Alan. (2006). *The Oxford Companion to Food.* 2nd ed. UK: Oxford University Press.

McGee, Harold. (2004).*On Food and Cooking: The Science and Lore of the Kitchen.* New York: Simon and Schuster.

Jango-Cohen, Judith. (2005). *The History of Food.* Twenty-First Century Books.

Mead, Margaret. (1997). *The Changing Significance of Food.* In Carole Counihan and Penny Van Esterik (Ed.), Food and Culture: A Reader. UK: Routledge.

Regmi, Anita (2001).*Changing Structure of Global Food Consumption and Trade.* Market and Trade Economics Division, Economic Research Service, USDA, May 30.

4

Patterns of Food Consumption

The growth in food demand and need is the result of the combined effects of world population growth to over 9 billion by 2050, rising incomes and dietary changes towards higher meat intake. Meat production is particularly demanding in terms of energy, cereal and water. Today, nearly half of the world's cereals are being used for animal feed.

Each day 200,000 more people are added to the world food demand. The world's human population has increased near fourfold in the past 100 years; it is projected to increase from 6.7 billion to 9.2 billion by 2050. It took only 12 years for the last billion to be added, a net increase of nearly 230,000 new people each day, who will need housing, food and other natural resources. The largest population increase is projected to occur in Asia, particularly in China, India and Southeast Asia, accounting for about 60% and more of the world's population by 2050. The rate of population growth, however, is still relatively high in Central America, and highest in Central and part of Western Africa. In relative numbers, Africa will experience the most rapid growth, over 70% faster than in Asia (annual growth of 2.4% versus 1.4% in Asia, compared to the global average of 1.3% and only 0.3% in many industrialized countries). In sub-Saharan Africa, the population is projected to increase from about 770 million to nearly 1.7 billion by 2050.

New estimates released by the World Bank in August 2008 show that in the developing world, the number of people living in extreme poverty may be higher than previously thought. With a threshold of extreme poverty set at US$1.25 a day, there were 1.4 billion people living in extreme poverty in 2005. Each year, nearly 10 million die of hunger and hunger-related diseases. While the proportion of underweight children below five years old decreased – from 33% in 1990 to 26% in 2006 – the number of children in developing countries who were underweight still exceeded 140 million. Similarly, while the proportion of impoverished persons might have declined in many regions, their absolute number has not fallen in some regions as populations continue to rise.

There are huge regional differences in the above trends. Globally, poverty rates have fallen from 52% in 1981 to 42% in 1990 and to 26% in 2005. In Sub-Saharan Africa, however, the poverty rate remained constant at around 50%. This region also comprises the majority of countries making the least progress in reducing child malnutrition. The poverty rate in East Asia fell from nearly 80% in 1980 to under 20% by 2005. East Asia, notably China, was successful in more than halving the proportion of underweight children between 1990 and 2006. In contrast, and despite improvements since 1990, almost 50% of the children are underweight in Southern Asia. This region alone accounts for more than half the world's malnourished children.

In addition to increasing demand for food by a rising population, observed dietary shifts also have implications for world food production. Along with rising population are the increasing incomes of a large fraction of the world's population. The result is increasing consumption of food per capita, as well as changes in diets towards a higher proportion of meat. With growing incomes, consumption – and quantity of waste or discarded food – increases substantially.

Diet Change

The global production of cereals (including wheat, rice and maize) plays a crucial role in the world food supply, accounting for about 50% of the calorie intake of humans. Any changes in the production of, or in the use of cereals for non-human consumption will have an immediate effect on the calorie intake of a large fraction of the world's population.

As nearly half of the world's cereal production is used to produce animal feed, the dietary proportion of meat has a major influence on global food demand. With meat consumption projected to increase from 37.4 kg/person/year in 2000 to over 52 kg/person/year by 2050, cereal requirements for more intensive meat production may increase substantially to more than 50% of total cereal production.

Promoting healthy diets and lifestyles to reduce the global burden of noncommunicable diseases requires a multisectoral approach involving the various relevant sectors in societies. The agriculture and food sector figures prominently in this enterprise and must be given due importance in any consideration of the promotion of healthy diets for individuals and population groups. Food strategies must not merely be directed at ensuring food security for all, but must also achieve the consumption of adequate quantities of safe and good quality foods that together make up a healthy diet.

Economic development is normally accompanied by improvements in a country's food supply and the gradual elimination of dietary deficiencies, thus improving the overall nutritional status of the country's population. Furthermore, it also brings about qualitative changes in the production, processing, distribution and marketing of food. Increasing urbanisation will also have consequences for the dietary patterns and lifestyles of individuals, not all of which are positive. Changes in diets, patterns of work and leisure — often referred to as the "nutrition transition" — are already contributing to the causal factors underlying noncommunicable diseases even in the poorest countries. Moreover, the pace of these changes seems to be accelerating, especially in the low-income and middle-income countries.

The dietary changes that characterise the "nutrition transition" include both quantitative and qualitative changes in the diet. The adverse dietary changes include shifts in the structure of the diet towards a higher energy density diet with a greater role for fat and added sugars in foods, greater saturated fat intake (mostly from animal sources), reduced intakes of complex carbohydrates and dietary fibre, and reduced fruit and vegetable intakes. These dietary changes are compounded by lifestyle changes that reflect reduced physical activity at work and during leisure time. At the same time, however, poor countries continue to face food shortages and nutrient inadequacies.

Diets evolve over time, being influenced by many factors and complex interactions. Income, prices, individual preferences and beliefs, cultural traditions, as well as geographical, environmental, social and economic factors all interact in a complex manner to shape dietary consumption patterns. Data on the national availability of the main food commodities provide a valuable insight into diets and their evolution over time. FAO produces annual Food Balance Sheets which provide national data on food availability (for almost all commodities and for nearly all countries).

Food Balance Sheets give a complete picture of supply (including production, imports, stock changes and exports) and utilisation (including final demand in the form of food use and industrial non-food use, intermediatedemandsuch as animal feed and seed use, and waste) by commodity. From these data, the average per capita supply of macronutrients (i. e. energy, protein, fats) can be derived for all food commodities. Although such average per capita supplies are derived from national data, they may not correspond to actual per capita availability, which is determined bymanyother factors such as inequality in access to food.

Likewise, these data refer to "average food available for consumption", which, for a number of reasons (for example, waste at the household level), is not equal to average food intake or average food consumption. Actual food availability may vary by region, socioeconomic level and season. Certain difficulties are encountered when estimating trade, production and stock changes on an annual scale. Hence three-year averages are calculated in order to reduce errors. The FAO statistical database (FAOSTAT), being based on national data, does not provide information on the distribution of food within countries, or within communities and households.

World Food Supply

The world food production has increased substantially in the past century, as has calorie intake per capita. However, in spite of a decrease in the proportion of undernourished people, the absolute number has in fact increased during the current food crisis, to over 963 million. By 2050, population growth by an estimated 3 billion more people will increase food demand.

Increased fertilizer application and more water usage through irrigation have been responsible for over 70% of the crop yield increase in the past. Yields, however, have nearly stabilized for cereals, partly as a result of low and declining investments in agriculture. In addition, fisheries landings have declined in the past decade mainly as a result of overfishing and unsustainable fishing methods.

Food supply, however, is not only a function of production, but also of energy efficiency. Food energy efficiency is our ability to minimize the loss of energy in food from harvest potential through processing to actual consumption and recycling. By optimizing this chain, food supply can increase with much less damage to the environment, similar to improvements in efficiency in the traditional energy sector. However, unlike the traditional energy sector, food energy efficiency has received little attention. Only an estimated 43% of the cereal produced is available for human consumption, as a result of harvest and post-harvest distribution losses and use of cereal for animal feed. Furthermore, the 30 million tonnes of fish needed to sustain the growth in aquaculture correspond to the amount of fish discarded at sea today.

A substantial share of the increasing food demand could be met by introducing food energy efficiency, such as recycling of waste. With new technology, waste along the human food supply chain could be used as a substitute for cereal in animal feed. The available cereal from such alternatives and efficiencies could feed all of the additional 3 billion people expected by 2050. At the same time, this would support a growing green economy and greatly reduce pressures on biodiversity and water resources – a truly 'win-win' solution.

Food from Crops

The three primary factors that affected recent increases in world crop production are:

1. Increased cropland and rangeland area (15% contribution in 1961–1999);
2. Increased yield per unit area (78% contribution); and
3. Greater cropping intensity (7% percent contribution).

The use of fertilizers accounts for approximately 50% of the yield increase, and greater irrigation for another substantial part. Current FAO projections in food demand suggest that cereal demand will increase by almost 50% towards 2050. This can either be obtained by increasing yields, continued expansion of cropland by conversion of natural habitats, or by optimizing food or feed energy efficiency from production to consumption.

Food from Fisheries and Aquaculture

Aquaculture, freshwater and marine fisheries supply about 10% of world human calorie intake – but this is likely to decline or at best stabilize in the future, and might have already reached the maximum. At present, marine capture fisheries yield 110–130 million tonnes of seafood annually. Of this,

70 million tonnes are directly consumed by humans, 30 million tonnes are discarded and 30 million tonnes converted to fishmeal.

The world's fisheries have steadily declined since the 1980s, its magnitude masked by the expansion of fishing into deeper and more offshore waters. Over half of the world's catches are caught in less than 7% of the oceans, in areas characterized by an increasing amount of habitat damage from bottom trawling, pollution and dead zones, invasive species infestations and vulnerability to climate change). Eutrophication from excessive inputs of phosphorous and nitrogen through sewage and agricultural run-off is a major threat to both freshwater and coastal marine fisheries. Areas of the coasts that are periodically starved of oxygen, so-called 'dead zones', often coincide with both high agricultural run-off and the primary fishing grounds for commercial and artisanal fisheries. Eutrophication combined with unsustainable fishing leads to the loss or depletion of these food resources, as occurs in the Gulf of Mexico, coastal China, the Pacific Northwest and many parts of the Atlantic, to mention a few.

Current projections for aquaculture suggest that previous growth is unlikely to be sustained in the future as a result of limits to the availability of wild marine fish for aquaculture feed. Small pelagic fish make up 37% of the total marine capture fisheries landings. Of this, 90% (or 27% of total landings) are processed into fishmeal and fish oil with the remaining 10% used directly for animal feed.

In some regions, such as in parts of Africa and Southeast Asia, increase in fisheries and expansion of cropland area have been the primary factors in increasing food supply. Indeed, fisheries are a major source of energy and protein for impoverished coastal populations, in particular in West Africa and Southeast Asia.Here, a decline in fisheries will have a major impact on the livelihoods and wellbeing of hundreds of millions of people.

Food from Meat

Meat production increased from 27 kg meat/capita in 1974/1976 to 36 kg meat/capita in 1997/1999, and now accounts for around 8% of the world calorie intake. In many regions, such as in the rangelands of Africa, in the Andes and the mountains of Central Asia, livestock is a primary factor in food security. Meat production, however, also has many detrimental effects on the environment, apart from being energy inefficient when animals are fed with food-crops. The area required for production of animal feed is approximately one-third of all arable land. Dietary shifts towards more meat will require a much larger share of cropland for grazing and feed production for the meat industry.

Expansion of land for livestock grazing is a key factor in deforestation, especially in Latin America: some 70% of previously forested land in the Amazon is used as pasture, with feed crops covering a large part of the remainder. About 70% of all grazing land in dry areas is considered degraded, mostly because of overgrazing, compaction and erosion attributable to livestock. Further, the livestock sector has an often unrecognized role in global warming – it is estimated to be responsible for 18% of greenhouse gas emissions, a bigger share than that of transport.

Food from Animal Feed

It takes, on average, 3 kg of grain to produce 1 kg of meat, given that part of the production is based on other sources of feed, rangeland and organic waste. Currently, 33 % of the cropland area is thus used for livestock. In addition, about 16,000 litres of virtual water are needed to produce 1 kg of meat. Hence, an increased demand for meat results in an accelerated demand for water, crop and rangeland area. Meat production is energy inefficient and environmentally harmful at industrial scales and with intense use of feed crops such as maize and soybeans. Chicken production is among the most energy-efficient, although still more energy-demanding than cereal production. Many farmers feed their animals organic waste from farm households or agricultural by-products that are unsuitable for human consumption. Small-scale pig farms often use organic residuals from restaurants and the food industry as fodder. If animals are part of an integrated farm production system, the overall energy efficiency can be actually increased through better utilization of organic waste. This is not the case for mass production of pigs and poultry in specialized stables, which may take up an increasingly larger proportion of the production of feed crops.

It is also important to note that much meat production takes place on extensive grasslands. But while often a threat to biodiversity and a source of competition with wild ungulates and birdlife, this requires very little or no input of commercial feed. Furthermore, it plays a crucial role in food security in many mountain areas, as well as in dry and steppe regions, including in Africa, Central Asia and the Andes.

Stabilizing the current meat production per capita by reducing meat consumption in the industrialized world and restraining it worldwide to 2000 level of 37,4 kg/capita in 2050 would free estimated 400 million tons of cereal per year for human consumption – or enough to cover the annual calorie need for 1.2 billion people in 2050. However, changing consumption patterns may be very difficult in the short-term. Increasing food supply by developing alternatives to cereals and improving feed efficiency in commercial feed may however have a much greater potential for increasing food supply.

Alternative Feed Sources

Choice of food – where choice exists – is a complex mix of traditions, religion, culture, availability and not the least, financial constraints. However, while many of these also apply to livestock, our ability to change the feed destined for livestock and aquaculture is probably greater than that of changing people's food choice habits, which are not as easily controlled. As cereal products are increasingly used as feed for livestock, estimated to be at least 35–40% of all cereal produced in 2008 and projected to reach nearly 45–50% by 2050 if meat consumption increases, finding alternative feed sources provides a huge potential for increasing the availability of cereal for human consumption. For other feed sources to become a sustainable alternative to the current use of cereals, their exploitation must not be resource-demanding. This poses a big challenge, since most of the easily available feed sources have already been fully exploited, although some alternatives still exist.

Cellulose is the most abundant biological material in the world, but the energy it contains is not readily available for animal production. Due to the interest in using this material for bioethanol production, there are currently large research programs underway to chemically and enzymatically degrade this cellulose into glucose. If this becomes possible and in a cost-effective manner, wood glucose can, to a large extent, replace cereals as a feed source for both ruminants and monogastric animals. Other fibrous plant sources such as straw, leaves and nutshells are also available in large quantities. Finding ways to feed the world's livestock is therefore a primary challenge.

Other sources for feed that are not fully exploited include seaweed, algae and other under-utilized marine organisms such as krill. However, their potential is uncertain, since technological challenges still remain. In addition, the impact of their harvesting on the ecosystem is of concern. The use of waste provides a much greater potential for alternative sources of animal feed.

Food or Feed from Waste

By using discards, waste and other post-harvest losses, the supply of animal and fish feed can be increased and be sustained without expanding current production, simply by increasing energy efficiency and conservation in the food supply chain.

There has been surprisingly little focus on salvaging food already harvested or produced. An important question centers around the percentage of food discarded or lost during harvesting, processing, transport and distribution as well as at the point of final sale to consumers. Reducing such losses is likely to be among the most sustainable alternatives for increasing food availability.

Discarded fish from marine fisheries is the single largest proportion lost of any food source produced or harvested from the wild. The proportion is particularly high for shrimp bottom trawl fisheries. Mortality among discarded fish is not adequately known, but has, for some species, been estimated to be as high as 70–80%, perhaps higher. Discarded fish alone amounts to as much as 30 million tonnes, compared to total landings of 100–130 tonnes/year. Feed for aquaculture is a major bottleneck, as there are limitations to the available oil and fish for aquaculture feed. A collapse in marine ecosystems would therefore have a direct impact on the prices of aquaculture products and on its scale of production. There is no indication that marine fisheries today can sustain the 23% increase in landings required for the 56% growth in aquaculture production required to maintain per capita fish consumption at current levels to 2050. However, if sustainable, the amount of fish currently discarded at sea could alone sustain more than a 50% increase in aquaculture production. However, many of these species could also be used directly for human consumption.

Fish post-harvest losses are generally high at the small-scale level. Recent work in Africa by FAO has shown that regardless of the type of fisheries (single or multi-species), physical post-harvest losses (that is, fish lost for human consumption) are commonly very low, typically around 5%. Downgrading of fish because of spoilage is considerable, however, perhaps as high as 10% and more. Hence, the total amount of fish lost through discards, post-harvest loss and spoilage may be around 40% of landings.The potential to use unexploited food waste as alternative sources of feed is also considerable for agricultural products.

Food losses in the field (between planting and harvesting) could be as high as 20–40% of the potential harvest in developing countries due to pests and pathogens. Postharvest losses vary greatly among commodities and production areas and seasons. In the United States, the losses of fresh fruits and vegetables have been estimated to range from 2% to 23%, depending on the commodity, with an overall average of about 12% losses between production and consumption sites. Kantor et al estimated the U.S. total retail, foodservice, and consumer food losses in 1995 to be 23% of fruits and 25% of vegetables. In addition, losses could amount to 25–50% of the total economic value because of reduced quality. Others estimate that up to 50% of the vegetables and fruits grown end as waste. Finally, substantial losses and wastage occur during retail and consumption due to product deterioration as well as to discarding of excess perishable products and unconsumed food. While the estimates therefore vary among sources, it is clear that food waste represents a major potential, especially for use as animal feed, which, in turn, could release the use of cereals in animal feed for human consumption.

In 2007, US$148 billion was invested in the renewable energy market, up 60% from the previous year. Recovering energy from agricultural wastes is becoming increasingly feasible at the industrial production level; investments in technology enhancement of existing systems and innovation in new waste management systems is called for to support this expanding green economy.

Developments in the Availability of Dietary Energy

Food consumption expressed in kilocalories (kcal) per capita per day is a key variable used for measuring and evaluating the evolution of the global and regional food situation. A more appropriate term for this variable would be "national average apparent food consumption" since the data come from national Food Balance Sheets rather than from food consumption surveys. Analysis of FAOSTAT data shows that dietary energy measured in kcals per capita per day has been steadily increasing on a worldwide basis; availability of calories per capita from the mid-1960s to the late 1990s increased globally by approximately 450 kcal per capita per day and by over

600 kcal per capita per day in developing countries (Table 1). This change has not, however, been equal across regions.

The per capita supply of calories has remained almost stagnant in sub-Saharan Africa and has recently fallen in the countries in economic transition. In contrast, the per capita supply of energy has risen dramatically in East Asia (by almost 1000 kcal per capita per day, mainly in China) and in the Near East/North Africa region (by over 700 kcal per capita per day). In short, it would appear that the world has made significant progress in raising food consumption per person. The increase in the world average consumption would have been higher but for the declines in the transition economies that occurred in the 1990s.

Table 1: Global and regional per capita food consumption (kcal per capita per day)

Region	1964-66	1974-76	1984-86	1997-99	2015	2030
World	2358	2435	2655	2803	2940	3050
Developing countries	2054	2152	2450	2681	2850	2980
Near East and North Africa	2290	2591	2953	3006	3090	3170
Sub-Saharan Africa	2058	2079	2057	2195	2360	2540
Latin America and the Caribbean	2393	2546	2689	2824	2980	3140
East Asia	1957	2105	2559	2921	3060	3190
South Asia	2017	1986	2205	2403	2700	2900
Industrialized countries	2947	3065	3206	3380	3440	3500
Transition countries	3222	3385	3379	2906	3060	3180

It is generally agreed, however, that those declines are likely to revert in the near future. The growth in food consumption has been accompanied by significant structural changes and a shift in diet away from staples such as roots and tubers towards more livestock products and vegetable oils. Table 1 shows that current energy intakes range from 2681 kcal per capita per day in developing countries, to 2906 kcal per capita per day in transition countries and 3380 kcal per capita per day in industrialised countries.

Globally, the share of dietary energy supplied by cereals appears to have remained relatively stable over time, representing about 50% of dietary energy supply. Recently, however, subtle changes appear to be taking place (Fig. 1). A closer analysis of the dietary energy intake shows a decrease in developing countries, where the share of energy derived from cereals has fallen from 60% to 54% in a period of only 10 years. Much of this downwards trend is attributable to cereals, particularly wheat and rice, becoming less preferred foods in middle-income countries such as Brazil and China, a pattern likely to continue over the next 30 years or so. Fig. 2 shows the structural changes in the diet of developing countries over the past 30—40 years and FAO's projections to the year 2030.

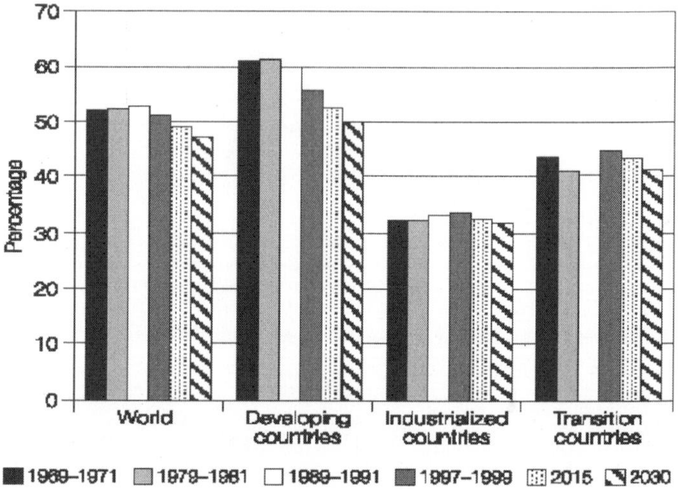

Figure 1: The Share of Dietary Energy Derived from Cereals

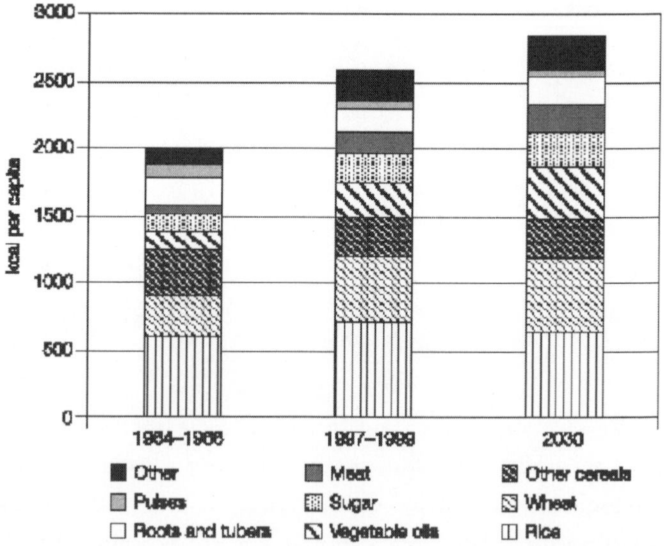

Figure 2: Calories from Major Commodities in Developing Countries

Consumption Patterns of Dietary Fat

The increase in the quantity and quality of the fats consumed in the diet is an important feature of nutrition transition reflected in the national diets of countries. There are large variations across the regions of the world in the amount of total fats (i. e. fats in foods, plus added fats and oils) available for human consumption. The lowest quantities consumed are recorded in Africa, while the highest consumption occurs in parts of North America and Europe. The important point is that there has been a remarkable increase in the intake of dietary fats over the past three decades and that this increase has taken place practically everywhere except in Africa, where consumption levels have stagnated.

The per capita supply of fat from animal foods has increased, respectively, by 14 and 4 g per capita in developing and industrialised countries, while there has been a decrease of 9 g per capita in transition countries. The increase in dietary fat supply worldwide exceeds the increase in dietary protein supply. The average global supply of fat has increased by 20 g per capita per day since 1967—1969. This increase in availability has been most pronounced in the Americas, East Asia, and the European Community. The proportion of energy contributed by dietary fats exceeds 30% in the industrialised regions, and in nearly all other regions this share is increasing.

The fat-to-energy ratio (FER) is defined as the percentage of energy derived from fat in the total supply of energy (in kcal). Country-specific analysis of FAO data for 1988—1990 found a range for the FER of 7—46%. Atotal of 19 countries fell below the minimum recommendation of15%dietary energy supply from fat, the majority of these being in sub-Saharan Africa and the remainder in South Asia. In contrast, 24 countries were above the maximum recommendation of 35%, the majority of these countries being in North America and Western Europe.

It is useful to note that limitations of the Food Balance Sheet data may contribute much of this variation in the FER between countries. For instance, in countries such as Malaysia with abundant availability of vegetable oils at low prices, Food Balance Sheet data may not reflect real consumption at the individual household level. Rising incomes in the developing world have also led to an increase in the availability and consumption of energy-dense high-fat diets. Food balance data can be used to examine the shift in the proportion of energy from fat over time and its relationship to increasing incomes.

In 1961-1963, a diet providing 20% of energy from fat was associated only with countries having at least a per capita gross national product of Rs. 66375. By 1990, however, even poor countries having a gross national product of only Rs. 33750 per capita had access to a similar diet comprising 20% of energy from fat. This change was mainly the result of an increase in the consumption of vegetable fats by poor countries, with smaller increases occurred in middle-income and high-income countries. By 1990, vegetable fats accounted for a greater proportion of dietary energy than animal fats for countries in the lowest per capita income category.

Changes in edible vegetable oil supply, in prices and in consumption equally affected rich and poor countries, although the net impact was relatively much greater in low-income countries. An equally large and important shift in the proportion of energy from added sugars in the diets of low-income countries was also a feature of the nutrition transition. Examinations of the purchasing habits of people, aimed at understanding the relationship between level of education or income and the different amounts or types of commodities purchased at different times were also revealing.

Research conducted in China shows that there have been profound shifts in purchasing practices in relation to income over the past decade. These analyses show how extra income in China affects poor people and rich people in a differential manner, enhancing the fat intake of the poor more than that of the rich. A variable proportion of these fat calories are provided by saturated fatty acids. Only in the two of the most affluent regions (i. e. in parts of North America and Europe) is the intake of saturated fat at or above 10% of energy intake level. In other less affluent regions, the proportion of dietary energy contributed by saturated fatty acids is lower, ranging from 5% to 8%, and generally not changing much over time.

National dietary surveys conducted in some countries confirm these data. The ratio of dietary fat from animal sources to total fat is a key indicator, since foods from animal sources are high in saturated fat. Data sets used to calculate country-specific FERs can also be used to calculate proportions of animal fat in total fat. Such analysis indicated that the proportion of animal fat in total fat was lower than 10% in some countries (Democratic Republic of Congo, Mozambique, Nigeria, Sao Tome and

Principe, and Sierra Leone), while it is above 75% in some other countries (Denmark, Finland, Hungary, Mongolia, Poland and Uruguay). These findings are not strictly divided along economic lines, as not all of the countries in the high range represent the most affluent countries. Country-specific food availability and cultural dietary preferences and norms to some extent determine these patterns.

The types of edible oils used in developing countries are also changing with the increasing use of hardened margarines (rich in trans fatty acids) that do not need to be refrigerated. Palm oil is becoming an increasingly important edible oil in the diets of much of South-East Asia and is likely to be a major source in the coming years.

Currently, palm oil consumption is low and the FER ranges between 15% and 18%. At this low level of consumption, the saturated fatty acid content of the diet comprises only 4%to 8%. Potential developments in the edible oil sector could affect all stages of the oil production process from plant breeding to processing methods, including the blending of oils aimed at producing edible oils that have a healthy fatty acid composition.

Olive oil is an important edible oil consumed largely in the Mediterranean region. Its production has been driven by rising demand, which has increasingly shifted olive cultivation from traditional farms to more intensive forms of cultivation. There is some concern that the intensive cultivation of olives may have adverse environmental impacts, such as soil erosion and desertification. However, agricultural production methods are being developed to ensure less harmful impacts on the environment.

Consumption Patterns of Animal Products

There has been an increasing pressure on the livestock sector to meet the growing demand for high-value animal protein. The world's livestock sector is growing at an unprecedented rate and the driving force behind this enormous surge is a combination of population growth, rising incomes and urbanisation. Annual meat production is projected to increase from 218 million tonnes in 1997—1999 to 376 million tonnes by 2030. There is a strong positive relationship between the level of income and the consumption of animal protein, with the consumption of meat, milk and eggs increasing at the expense of staple foods. Because of the recent steep decline in prices, developing countries are embarking on higher meat consumption at much lower levels of gross domestic product than the industrialised countries did some 20—30 years ago.

Urbanisation is a major driving force influencing global demand for livestock products. Urbanisation stimulates improvements in infrastructure, including cold chains, which permit trade in perishable goods. Compared with the less diversified diets of the rural communities, city dwellers have a varied diet rich in animal proteins and fats, and characterised by higher consumption of meat, poultry, milk and other dairy products.

There has been a remarkable increase in the consumption of animal products in countries such as Brazil and China, although the levels are still well below the levels of consumption in North American and most other industrialised countries. As diets become richer and more diverse, the high-value protein that the livestock sector offers improves the nutrition of the vast majority of the world. Livestock products not only provide high-value protein but are also important sources of a wide range of essential micronutrients, in particular minerals such as iron and zinc, and vitamins such as vitamin A.

For the large majority of people in the world, particularly in developing countries, livestock products remain a desired food for nutritional value and taste. Excessive consumption of animal products in some countries and social classes can, however, lead to excessive intakes of fat. The growing demand for livestock products is likely to have an undesirable impact on the environment. For example, there will be more large-scale, industrial production, often located close to urban centres, which brings with it a range of environmental and public health risks. Attempts have been made to estimate the environmental impact of industrial livestock production.

For instance, it has been estimated that the number of people fed in a year per hectare ranges from 22 for potatoes and 19 for rice to 1 and 2, respectively, for beef and lamb. The low energy conversion ratio from feed to meat is another concern, since some of the cereal grain food produced is diverted to livestock production. Likewise, land and water requirements for meat production are likely to become a major concern, as the increasing demand for animal products results in more intensive livestock production systems.

Consumption of Fish

Despite fluctuations in supply and demand caused by the changing state of fisheries resources, the economic climate and environmental conditions, fisheries, including aquaculture, have traditionally been, and remain an important source of food, employment and revenue in many countries and communities. After the remarkable increase in both marine and inland capture of fish during the 1950s and 1960s, world fisheries production has levelled off since the 1970s. This levelling off of the total catch follows the general trend of most of the world's fishing areas, which have apparently reached their maximum potential for fisheries production, with the majority of stocks being fully exploited. It is therefore very unlikely that substantial increases in total catch will be obtained in the future.

In contrast, aquaculture production has followed the opposite path. Starting from an insignificant total production, inland and marine aquaculture production has been growing at a remarkable rate, offsetting part of the reduction in the ocean catch of fish. The total food fish supply and hence consumption has been growing at a rate of 3. 6% per year since 1961, while the world's population has been expanding at 1. 8% per year. The proteins derived from fish, crustaceans and molluscs account for between 13. 8% and 16. 5% of the animal protein intake of thehumanpopulation.

The average apparent per capita consumption increased from about 9 kg per year in the early 1960s to 16 kg in 1997. The per capita availability of fish and fishery products has therefore nearly doubled in 40 years, outpacing population growth. As well as income-related variations, the role of fish in nutrition shows marked continental, regional and national differences. In industrialised countries, where diets generally contain a more diversified range of animal proteins, a rise in per capita provision from 19. 7 kg to 27. 7 kg seems to have occurred. This represents a growth rate close to1%per year. In this group of countries, fish contributed an increasing share of total protein intake until 1989 (accounting for between 6. 5% and 8. 5%), but since then its importance has gradually declined and, in 1997, its percentage contribution was back to the level prevailing in the mid-1980s.

In the early 1960s, per capita fish supply in low-income food-deficit countries was, on average, only 30% of that of the richest countries. This gap has been gradually reduced, such that in 1997, average fish consumption in these countries was 70% of that of the more affluent economies. Despite the relatively low consumption by weight in low-income food-deficit countries, the contribution of fish to total animal protein intake is considerable (nearly 20%). Over the past four decades, however, the share of fish proteins in animal proteins has declined slightly, because of faster growth in the consumption of other animal products. Currently, two-thirds of the total food fish supply is obtained from capture fisheries in marine and inland waters, while the remaining onethird is derived from aquaculture.

The contribution of inland and marine capture fisheries to per capita food supply has stabilised, around 10 kg per capita in the period 1984—1998. Any recent increases in per capita availability have, therefore, been obtained from aquaculture production, from both traditional rural aquaculture and intensive commercial aquaculture of high-value species. Fish contributes up to 180 kcal per capita per day, but reaches such high levels only in a few countries where there is a lack of alternative protein foods grown locally or where there is a strong preference for fish (examples are Iceland, Japan and some small island states).

More typically, fish provides about 20—30 kcal per capita per day. Fish proteins are essential in the diet of some densely populated countries where the total protein intake level is low, and are very important in the diets of many other countries. Worldwide, about a billion people rely on fish as their main source of animal proteins. Dependence on fish is usually higher in coastal than in inland areas. About 20% of the world's population derives at least one-fifth of its animal protein intake from fish, and some small island states depend almost exclusively on fish.

Recommending the increased consumption of fish is another area where the feasibility of dietary recommendations needs to be balanced against concerns for sustainability of marine stocks and the potential depletion of this important marine source of high quality nutritious food. Added to this is the concern that a significant proportion of the world fish catch is transformed into fish meal and used as animal feed in industrial livestock production and thus is not available for human consumption.

Consumption Patterns of Fruits and Vegetables

Consumption of fruits and vegetables plays a vital role in providing a diversified and nutritious diet. Alow consumption of fruits and vegetables in many regions of the developing world is, however, a persistent phenomenon, confirmed by the findings of food consumption surveys. Nationally representative surveys in India, for example, indicate a steady level of consumption of only 120—140 g per capita per day, with about another 100 g per capita coming from roots and tubers, and some 40 g per capita from pulses. This may not be true for urban populations in India, who have rising incomes and greater access to a diverse and varied diet. In contrast, in China, — a country that is undergoing rapid economic growth and transition — the amount of fruits and vegetables consumed has increased to 369 g per capita per day by 1992.

At present, only a small and negligible minority of the world's population consumes the generally recommended high average intake of fruits and vegetables. In 1998, only 6 of the 14 WHO regions had an availability of fruits and vegetables equal to or greater than the earlier recommended intake of 400 g per capita per day. The relatively favourable situation in 1998 appears to have evolved from a markedly less favourable position in previous years, as evidenced by the great increase in vegetable availability recorded between 1990 and 1998 for most of the regions. In contrast, the availability of fruit generally decreased between 1990 and 1998 in most regions of the world. The increase in urbanisation globally is another challenge.

Increasing urbanisation will distance more people from primary food production, and in turn have a negative impact on both the availability of a varied and nutritious diet with enough fruits and vegetables, and the access of the urban poor to such a diet. Nevertheless, it may facilitate the achievement of other goals, as those who can afford it can have better access to a diverse and varied diet. Investment in periurban horticulture may provide an opportunity to increase the availability and consumption of a healthy diet. Global trends in the production and supply of vegetables indicate that the current production and consumption vary widely among regions. It should be noted that the production of wild and indigenous vegetables is not taken into account in production statistics and might therefore be underestimated in consumption statistics.

In 2000, the global annual average per capita vegetable supply was 102 kg, with the highest level in Asia (116 kg), and the lowest levels in South America (48 kg) and Africa (52 kg). These figures also include the large amount of horticultural produce that is consumed on the farm.

From Supply to Food Security

Food security is not simply a function of production or supply, but of availability, accessibility, stability of supply, affordability and the quality and safety of food. These factors include a broad spectrum of socio-economic issues with great influence on farmers and on the impoverished in particular.Large shares of the world's small-scale farmers, particularly in central Asia and in Africa, are constrained by access to markets, while inputs, such as fertilizers and seed, are expensive. With lack of irrigation water, infrastructure and investments, and low availability of micro-finance combined with dependency on few multinational suppliers, crop production is unlikely to increase in those regions where it is needed the most, unless major policy changes and investments take place. These constraints are further compounded by conflicts and corruption.

Agricultural prices are forecast to decline over the next two years but to remain well above the levels of the first half of this decade. However, by 2030–2050, the current scenarios of losses and constraints due to climate change and environmental degradation – with no policy change – suggest that production increases could fall to 0.87% towards 2030 and to 0.5% by 2030–2050. Should global agricultural productivity rise by less than 1.2% per year on average, then prices, rather than declining, can be expected to rise by as much as 0.3% per year. In addition, a production short of demand, a greater geographical inequity in production and demand, combined with possibly more extreme weather and subsequent speculation in food markets, could generate much greater price volatility than before. In turn, this could potentially induce a substantially greater reduction in food security than that seen in the current crisis, if appropriate options for increasing supply and security are not considered and implemented.

Whether the Millennium Development Goals (MDGs) like hunger eradication will be met in the (near) future and whether the food crisis as evolved until 2008 will impact these MDGs on the longer term, depends on how markets will respond, how price impacts will cascade through the food production system and how international governments will respond to these new circumstances. In short, the impact on food availability and food security can only be assessed through the different dimensions that play a role in the state of food security. The FAO defines food security as follows: "Food security exists when all people, at all times, have physical and economic access to sufficient, safe and nutritious food for a healthy and active life". This involves four dimensions:

— Adequacy of food supply or availability;
— Stability of supply, without seasonal fluctuations or shortages;

— Accessibility to food or affordability; and

— Utilization: quality and safety of food.

Before conclusions can be drawn on food security, these dimensions need to be examined. The first three dimensions are elaborated upon in this chapter. The fourth dimension of food utilization is beyond the scope of this report, of which the focus is the environmental aspects of food security.

The availability of food within a specific country can be guaranteed in two ways: Either by food production in the country itself or by trade. The first option has been discussed extensively in the previous chapters. The second option has become more and more important), with increasing transport possibilities and storing capacities and the growing challenges faced by some countries in their domestic production, including because of limitations in available cropland. International trade in agricultural products has expanded more rapidly than global agricultural GDP.

The past several decades have witnessed a major increase in the integration of the world economy through trade. Many parts of the world have experienced high economic growth in recent years. For example, Asia's GDP has increased by 9% annually between 2004 and 2006, and growth is especially high in China and India. Sub-Saharan Africa experienced 6% annual growth in the same period, after a long period of recession in many countries. Even countries with a prevalence of hunger reported some economic growth, although this is not always reflected in social conditions. However, global economic growth is projected to slow to around 4% and be in the 6% range for developing countries beyond 2008.

An increasing share of global agricultural exports originates from developed countries. It increased from 32% in 2000 to 37% in 2006, but there are large regional variations. For instance, Africa's share in global exports only increased from 2.3 to 2.8% in this period. The EU countries account for most of the global growth; their share of total agricultural exports has increased from slightly more than 20% in the early 1960s to more than 40% today.

A large portion of this increase is accounted for by intra-EU trade, which represents around 30% of world agricultural trade. Conversely, during the past four decades, developing countries have seen their share of world agricultural exports decline from almost 40% to around 25% in the early 1990s before rebounding to about 30% today. This contrasts with the steadily increasing share of developing countries in total merchandise exports. Over this same period, the share of global agricultural imports purchased by developing countries increased from less than 20% to about 30%.

Another perspective of this trade is the purchase of land abroad for food production. Responding to recent food crises, a number of countries have started to purchase land abroad for cultivation of crops needed to support domestic demand. This is seen as a long-term solution to the high prices of agriculture commodities and increasing demand for agroforestry products such as palm oil. Among the most active countries owning, leasing or concessioning farmland overseas are China, India, Japan, Saudi Arabia, South Korea and United Arab Emirates; a number of other countries are only starting negotiations for the coming years. The total area of overseas farmland in different countries was estimated at 5.7 million ha at the end of 2008 or 0.4% of the global cropland area.

Increasing Food Production

Another option for meeting food demand is to ensure production in the country or region itself, by aiming at self-sufficiency and lowering the dependency on other regions. Current estimates of the developments on the demand side require increase in production in those regions with the highest economic growth or population increase. The majority of these regions will be in emerging economies in Africa and Asia. Nowadays, Africa is especially dependent on food imports. Food production in this region is lagging behind due to limited research investments and the problems for farmers to use the appropriate inputs in their production process.

Research Investments

The world regions are sharply divided in terms of their capacity to use science in promoting agricultural productivity in order to achieve food security and reduce poverty and hunger. For every US$100 of agricultural output, developed countries spend US$2.16 on public agricultural research and development (R&D), whereas developing countries spend only US$0.55. Total agricultural R&D spending in developing countries increased from US$3.7 billion to US$4.4 billion, or by 1.6% annually. This spending was largely driven by Asia, where annual spending increased by 3.3 percent. Today, Asia accounts for 42% of total agricultural R&D spending in developing countries (with China and India accounting for 18 and 10%, respectively). In Africa, agricultural R&D expenditure declined slightly, by 0.4%/year. Although Africa is geographically large, its share in R&D spending is only 13%. Latin America accounts for 33% (with Brazil being responsible for 48% of the region's spending).

Productivity has risen in many developing countries, mainly as a result of investment in agricultural R&D combined with improved human capital and rural infrastructure. In East Asia, land productivity increased from US$1,485/ha in 1992 to US$2,129/ha in 2006, while labour productivity rose from US$510 to US$822/worker. In Africa, the levels of productivity are much lower and their growth has also been slower. In 1992, land productivity in Sub-Saharan Africa was only 79% of that in East Asia; by 2006 this gap of 21% had increased to 59%.

Resources for Fertilizer Use

One of the major options for significantly raising crop production is increasing the use of mineral fertilizers. The Africa Fertilizer Summit 2006 concluded that the use of fertilizers should be increased to a level of at least 50 kg/ha by 2015. The present use of fertilizers in Sub-Saharan Africa is only about 9 kg/ha of arable land, compared to a world average of 101 kg/ha. Within Africa, there are strong differences in fertilizer use between regions, with relatively high use in Northern and Southern Africa, and very low use (around 1 to 2 kg/ha) in Western and Central Africa. Taking the increase as proposed by the Africa Fertilizer Summit as a starting point, this would mean a growth of the yearly use of fertilizers from 1 to 6 million tonnes. Based on the price of fertilizer (DAP) of approximately US$600/tonne, this would mean US$3 billion/year for the purchase of DAP only. A more moderate price of US$200/tonne would still mean US$1 billion/year. Added to this are significant costs of and investments in transport and distribution, developing agricultural research, extension programs, capacity building, etc. Indeed, there are many reasons for this low use. One of the reasons is the high retail prices of fertilizers, especially in areas with poor infrastructure. A metric tonne of urea costs $90 in Europe, $120 kg in the harbor of Mombassa, $400 in Western Kenya and $770 in Malawi.

A major challenge is to find ways of making fertilizer available to smallholders at affordable prices. There is also a need for holistic approaches to soil fertility management that embraces the full range of driving factors and consequences of soil degradation. This would include the integration of mineral and organic sources of nutrients, thereby using locally available sources of inputs and maximizing their use efficiency, while reducing dependency upon prices of commercial fertilizers and pesticides. The use of perennials, intercropping and agroforestry systems, such as the use of nitrogen fixating leguminous trees, are ways to increase nutrient availability, but also enhance water availability and pest control, in a more sustainable manner.

A major challenge is to find ways of making fertilizer available to smallholders at affordable prices. There is also a need for holistic approaches to soil fertility management that embraces the full range of driving factors and consequences of soil degradation. This would include the integration of mineral and organic sources of nutrients, thereby using locally available sources of inputs and maximizing their use efficiency.

Resources for Irrigation

Irrigated land area increased rapidly until 1980 with expansion rates of more than 2% a year. In Asia in particular, it led to a steady increase of staple food production together with other elements of the green revolution package. After 1980, growth in expansion of irrigated area decreased and it is assumed this trend will continue in the near future. One of the reasons is that the areas most suitable for irrigation are already used, leading to higher construction costs in new areas. Another reason is the strong decline in relative food prices over the last decades, which makes it less profitable to invest in irrigation. Current irrigation systems could be improved by investing in water control and delivery, automation, monitoring and staff training.

The irrigated area has remained very low in Sub-Saharan Africa and of the land under irrigation, 18% is not used. In most African regions the major challenge is not the lack of water, but unpredictable and highly variable rainfall patterns with occurrences of dry spells every two years causing crop failure. This high uncertainty and variability drive the risk-averse behaviour of smallholder farmers. Rarely are investments made in soil management and fertility, crop varieties, tillage practices and even labour in order to avoid losses in case of total crop failure. Managing the extreme rainfall variability over time and space can provide supplemental irrigation water to overcome dry periods and prevent crop failure. In combination with improved soil management (in regions with severe land degradation, only

5% of the rainwater is used for crops), this should reduce the risk of total crop failure and enhance the profitability of investments in crop management, for example, fertilizers, labour and crop varieties. Increasing crop canopy coverage reduces evapo-transpiration from the soil, improving soil moisture and the provision of water for the crop.

Stability of Food Supply

The second dimension of food security is the stability of food supply. Temporary disruption of supplies can have long-term impacts. The two options for fulfilling demand - food imports and domestic production - imply several reasons for instability of food supplies. A major reason for instability in food supply is high fluctuation in food prices (price volatility). Volatile prices lead to poor investment strategies of producers and immediate impacts on consumers, especially in developing countries where consumers spend a large share of their income on food. Another source of instability is conflicts, which increase food supply risks.

Price Volatility

Low and fluctuating prices are a core problem for stable food production. Agricultural price volatility increases the uncertainty faced by farmers and affects their investment decisions, productivity and income. Lagging investments can be a constraint in meeting changing consumer demands. For willingness to invest it is the volatility of the revenue flows that matters. Instability in prices is related to factors in the agricultural domain as well as in other sectors.

Trade policies that limit market access, increase the volatility of commodity prices, unfairly subsidize developed country exports and constrain the trade policy flexibility of the developing world affect the stability and security as well as overall economic wellbeing of developing countries. A quarter of the world's governments implemented some export restrictions in the current period of high prices to ensure domestic food security. The impacts of these restrictions varied from panic-buying to the cultivation of smaller areas due to high input costs and the expectation of low product prices. These restrictions even increased price volatility of food products on the world market, thereby decreasing the food security of other countries. Earlier experience shows that attempts to gain domestic price stability create global price instability. Furthermore, once policies are established to protect food markets, they are not easily dismantled.

It should also be noted that global food prices are determined by a small share of food products that are traded on the global market. The share of cereals traded compared to the volume produced is small and has increased slightly over the last four decades, from 9% to 13%. Annual fluctuations in world cereal production are in the same order of magnitude, varying from +9.8% to -3.9% of the previous year's production. This implies that supplies to the world market (the sum of the surplus in the supply of each region) can be reduced by one-third or increase two-fold. Demand in the world market does not follow this trend, however, and probably even moves in the opposite direction in case of poor harvests. These yearly trends describe the risk of discrepancy between supply and demand on the world food market. For this reason, with open markets, developing countries are very vulnerable to fluctuations in global food supply and prices and temporary protection of their own agricultural markets is promoted for these countries.

Supplies from food stocks can also buffer shortages on the world market. Stocks of cereals and vegetable oil have fallen to low levels relative to use, reducing the buffer against shocks in supply and demand. Stocks are not expected to be fully replenished over the coming 10 years, implying that tight markets may be a permanent factor in the next decade. This does not necessarily lead to permanent higher prices, but provides the background for continuing price volatility in the future.

Constraints by Conflicts

Conflicts increase the risk of food supply instability tremendously. Countries in conflict and post-conflict situations tend to be food insecure, with more than 20% of the population, and in many cases far more, lacking access to adequate food. The group of countries that are experiencing civil conflicts cannot meet their basic needs and are large importers of food. In addition, the transport of commodities is hazardous and the situation is not secure enough for farmers to make investment decisions.

Accessibility to Food

Accessibility to food concerns both physical access and affordability. Access to markets concerns transportation of commodities and its costs as well as the transmission of price developments to producers. Poor transmission of price incentives to producers results in broadening the gap between consumers and producers, especially in periods of changing diets.

Access to Markets

According to the latest UN estimates, almost all of the world's population growth between 2000 and 2030 will be concentrated in urban areas in developing countries. By 2030, almost 60% of the people in developing countries will live in cities. If present trends continue, urban population will equal rural population by around 2017.

Large urban markets create the scope for the establishment of big supermarket chains, with implications for the entire food supply chain. In 2002, the share of supermarkets in the processed/packaged food retail market was 33% in Southeast Asia and 63% in East Asia. The share of supermarkets in the fresh foods market was roughly 15-20% in Southeast Asia and 30% in East Asia outside of China. The 2001 supermarket share of Chinese urban food markets was 48%, up from 30% in 1999. Supermarkets are also becoming an emerging force in South Asia, particularly in urban India since the mid-1990s.

The increasing growth and power of international food corporations are affecting the opportunities of small agricultural producers in developing countries. While new opportunities are being created, the majority are not able to utilize them because of the stringent safety and quality standards of food retailers, hence barring market entry. The economy of the corporate food supply chain has grown steadily over the past years. Between 2004 and 2006 total global food spending grew by 16% from US$5.5 to US$6.4 trillion. In this period, the sales of food retailers increased disproportionately to the sales of food processors and companies in the food input industry. The sales of top food processors and traders grew by 13%, sales by the top 10 agricultural input countries by 8%, while the sales by top food retailers grew by 40%. However, on a global scale the agricultural input industry is more monopolized than the food retail industry. In the agricultural input industry, three agro-chemical corporations control approximately half the world market, while the top five food retailers control only around 13% of the total market.

Trade and urbanization affect consumer preferences. The rapid diversification of the urban diet cannot be met by the traditional food supply chain in the hinterland of many developing countries. Consequently, importing food to satisfy the changing food demand could be relatively easier and less costly than acquiring the same food from domestic sources.

In Asia traditional rice-eating societies are consuming increasing quantities of wheat in the form of bread, cakes, pastry and other products. Countries that traditionally imported rice for meeting food shortfalls may now be shifting towards increasing levels of wheat imports. This trend is also evident in the import of other temperate products like vegetables, milk and dairy products and temperate fruit. Net imports of this category of products increased by a factor of 13 over the last 40 years, rising from a deficit of US$1.7 billion in 1961/1963 to US$24 billion in 1997. Between 1997/1999 and 2030, the cumulative increase in imports of these products is expected to be 154% and 17% for vegetable oils and oilseeds, while meat imports are expected to increase by 389%. The overall result is that we are beginning to see a homogenization of food tastes across the globe, but with regional variations.

Poor connections between urban and rural areas hinder price transmissions towards local markets, broadening the gap between urban demand and rural production in increasing demand for traditional products or for product diversification. The lack of access to markets is most evident in Africa, although large parts of Latin America and Asia are also experiencing long transport hours to reach markets. Consequently, domestic prices do not always follow international prices. The periods of rising real prices were generally associated with real exchange rate devaluations. Relaxation of government controls over prices and market systems also led to gains in producer prices in some cases. In other instances, import liberalization appears to have contributed to a decline in the real domestic prices of some commodities. Consequently, global shortages of food and feed that lead to global price increases are not followed by production increases at the local level.

Future Trends in Demand, Availability and Consumption of Food

In recent years the growth rates of world agricultural production and crop yields have slowed. This has raised fears that the world may not be able to grow enough food and other commodities to ensure that future populations are adequately fed. However, the slowdown has occurred not because of shortages of land or water but rather because demand for agricultural products has also slowed. This is mainly because world population growth rates have been declining since the late 1960s, and fairly high levels of food consumption per person are now being reached in many countries, beyond which further rises will be limited.

It also true that a high share of the world's population remains in poverty and hence lacks the necessary income to translate its needs into effective demand. As a result, the growth in world demand for agricultural products is expected to fall from an average 2. 2% per year over the past 30 years to an average 1. 5% per year for the next 30 years. In developing countries the slowdownwill be more dramatic, from 3. 7% per year to 2% per year, partly as a result of China having passed the phase of rapid growth in its demand for food. Global food shortages are unlikely, but serious problems already exist at national and local levels, and may worsen unless focused efforts are made. The annual growth rate of world demand for cereals has declined from 2. 5% per year in the 1970s and 1. 9% per year in the 1980s to only1%per year in the 1990s.

Annual cereal use per person (including animal feeds) peaked in the mid-1980s at 334 kg and has since fallen to 317 kg. The decline is not a cause for alarm, it is largely the natural result of slower population growth and shifts in human diets and animal feeds. During the 1990s, however, the decline was accentuated by a number of temporary factors, including serious economic recessions in the transition countries and in some East and South-East Asian countries. The growth rate in the demand for cereals is expected to rise again to 1. 4% per year up until 2015, slowing to 1. 2% per year thereafter. In developing countries overall, cereal production is not expected to keep pace with demand.

The net cereal deficits of these countries, which amounted to 103 million tonnes or 9% of consumption in 1997-1999, could rise to 265 million tonnes by 2030, when they will be 14% of consumption. This gap can be bridged by increased surpluses from traditional grain exporters, and by new exports from the transition countries, which are expected to shift from being net importers to being net exporters. Oil crops have seen the fastest increase in area of any crop sector, expanding by 75 million hectares between the mid-1970s and the end of the 1990s, while cereal area fell by 28 million hectares over the same period.

Future per capita consumption of oil crops is expected to rise more rapidly than that of cereals. These crops will account for 45 out of every 100 extra kilocalories added to average diets in developing countries between now and 2030. There are three main sources of growth in crop production: expanding the land area, increasing the frequency at which it is cropped (often through irrigation), and boosting yields. It has been suggested that growth in crop production may be approaching the ceiling of what is possible in respect of all three sources.

Diets in developing countries are changing as incomes rise. The share of staples, such as cereals, roots and tubers, is declining, while that of meat, dairy products and oil crops is rising. Between 1964-1966 and 1997-1999, per capita meat consumption in developing countries rose by 150% and that of milk and dairy products by 60%. By 2030, per capita consumption of livestock products could rise by a further 44%. Poultry consumption is predicted to grow the fastest. Productivity improvements are likely to be a major source of growth.

Milk yields should improve, while breeding and improved management should increase average carcass weights and off-take rates. This will allow increased production with lower growth in animal numbers, and a corresponding slowdown in the growth of environmental damage from grazing and animal wastes. In developing countries, demand is predicted to grow faster than production, resulting in a growing trade deficit. In meat products this deficit will rise steeply, from 1.2 million tonnes per year in 1997-1999 to 5.9 million tonnes per year in 2030 (despite growing meat exports from Latin America), while in the case of milk and dairy products, the rise will be less steep but still considerable, from 20 million tonnes per year in 1997-1999 to 39 million tonnes per year in 2030.

An increasing share of livestock production will probably come from industrial enterprises. In recent years, production from this sector has grown twice as fast as that from more traditional mixed farming systems and more than six times faster than that from grazing systems. World fisheries production has kept ahead of population growth over the past three decades. Total fish production has almost doubled, from 65 million tonnes in 1970 to 125 million tonnes in 1999, when the world average intake of fish, crustaceans and molluscs reached 16. 3 kg per person. By 2030, annual fish consumption is likely to rise to some 150-160 million tonnes, or between 19-20 kg per person. This amount is significantly lower than the potential demand, as environmental factors are expected to limit supply.

During the 1990s the marine catch levelled out at 80-85 million tonnes per year, and by the turn of the century, three-quarters of ocean fish stocks were overfished, depleted or exploited up to their maximum sustainable yield. Further growth in the marine catch can only be modest. Aquaculture compensated for this marine slowdown, doubling its share of world fish production during the 1990s. It is expected to continue to grow rapidly, at rates of 5-7% per year up to 2015. In all sectors of fishing it will be essential to pursue forms of management conducive to sustainable exploitation, especially for resources under common ownership or no ownership.

Future World Food Prices

Accessibility to food is also determined by the long-term trend in food prices (which is a different issue from price volatility). The rising trend in global food prices is likely to persist in the next decade. In the long run, however, prices will decline.

Prices are driven by a complex combination of factors. Historically, productivity gains and increasing competition in trade have overtaken stronger demand, resulting in the declining trend of the past 100 years. Recently, food prices have been driven by a combination of rising fuel costs, production of biofuels, and unfavorable weather conditions, with trade restrictions boosting upward price pressures.

Agricultural prices are forecast to decline over the next two years, but remain well above the levels of the first half of this decade. A strong combination of supply response and continued growth in demand is expected to keep prices above historical levels, but well below the peaks experienced in recent years. In real terms, prices in 2017 are projected to be 10% to 35% higher than in the past decade. While the long-term outlook for agricultural prices is particularly uncertain, the decline is expected to continue. In Global Economic Prospects, the World Bank projects a decline of about 0.7% a year through the forecast period until 2030.

Price projections depend on a wide range of factors, including climate change, productivity developments, GDP and population growth and the policy environment. One of the cruel ironies today is seen in the connection between rising energy and food prices. Higher energy prices have increased fertilizer and transport costs and stimulated biofuel production (see box). This coupling can have devastating implications for global poverty and food security.

The impact of climate change is a particularly difficult issue in the long-term forecasts for agricultural prices. Forecasts of the rise in temperature and its impact on agriculture over the next two decades are extremely uncertain. Climate change threatens yields in many developing countries, although most of this effect is not likely to be felt until after 2030. The World Bank assumes an overall decline in agricultural productivity of between 1-10% by 2030 (compared with a future where average global temperatures remain stable), with Canada and Europe least affected and India, Sub-Saharan Africa, and parts of Latin America most affected. Were there to be no climate change between now and 2030, global agricultural productivity would be nearly 4% higher and the world price of food 5.3% lower. Over the longer term, the impacts of climate change could be much more serious, with agricultural productivity in many developing regions, notably Africa, potentially declining much higher than the global average.

In Global Economic Prospects 2009, the World Bank has run a number of simulations to quantify possible outcomes. Should global agricultural productivity rise by only 1.2% per year on average, instead of the 2.1% projected in the baseline, then prices, rather than declining, can be expected to rise by as much as 0.3% per year. If cereal production could increase as projected without any environmental constraints, it is expected to grow by 1.5-1.5% to 2030 according to demand, and by 0.9-1.0% between 2030 and 2050. However, the current scenarios of losses and constraints due to climate change and environmental degradation - with no policy change - suggest that production increases could fall to 0.87% towards 2030 and only 0.5% between 2030-205.

Alternatively, biofuels could have a significant impact on food prices if oil prices remain high or the cost of biofuels production declines. With a permanent increase in the rate of growth of demand for food products as source material for biofuels (assuming a doubling in biofuel production compared to the baseline), food prices will decline by only 0.5% a year. In general, OECD-FAO estimates confirm this sensitivity to key assumptions about yield and biofuels production.

Overall, soaring food prices are blamed for their impacts on human vulnerability. However, there are two sides to this picture. Increasing food prices do have a positive effect on net food-selling households, augmenting their incomes and allowing more possibilities for farmers to afford investments in production inputs. This underlines the need to minimize short-term price volatility and stimulate slow increases in long-term food prices, in order to enhance investments in the agricultural system and bridge the gap between developed and developing countries as well as between rural food producing and urban food consuming regions. Ideally, these developments should take the environmental aspects previously described into account to achieve sustainable agricultural systems that will meet the food demand of all the world citizens and eradicate hunger. However, increasing yield and food supply without simply continuing the conventional expansion of cropland and rangeland and use of fertilizers and pesticides - at the cost of biodiversity and future generations - will require major investments and implementation of food energy considerations in the entire food production and consumption chain.

References

Campbell, Bernard Grant. (1998). *Human Evolution: An Introduction to Man's Adaptations*. Aldine Transaction:

Carpenter, Ruth Ann; Finley, Carrie E. (2005). *Healthy Eating Every Day*. Human Kinetics.

Food and Agriculture Organization of the United Nations. *The State of Food Insecurity in the World 2005*. . Retrieved fromhttp://www.fao.org/docrep/008/a0200e/a0200e00.htm on 2006-09-29.

5

Methods of Cooking

Cooking is the process of preparing food, by the analog skills, often with the use of heat. Cooking techniques and ingredients vary widely across the world, reflecting unique environmental, economic, and cultural traditions. Cooks themselves also vary widely in skill and training. Cooking can also occur through chemical reactions without the presence of heat, most notably as in Ceviche, a traditional South American dish where fish is cooked with the acids in lemon or lime juice. Sushi also utilizes a similar chemical reaction between fish and the acidic content of rice glazed with vinegar.

Preparing food with heat or fire is an activity unique to humans, and some scientists believe the advent of cooking played an important role in human evolution. Most anthropologists believe that cooking fires first developed around 250,000 years ago. The development of agriculture, commerce and transportation between civilizations in different regions offered cooks many new ingredients. New inventions and technologies, such as pottery for holding and boiling water, expanded cooking techniques. Some modern cooks apply advanced scientific techniques to food preparation.

History of Cooking

There is no clear archaeological evidence when food was first cooked. Most anthropologists believe that cooking fires began only about 250,000 years ago, when hearths started appearing. Phylogenetic analysis by Chris Organ, Charles Nunn, Zarin Machanda, and Richard Wrangham suggests that cooking may have been invented as far back as 1.8 million to 2.3 million years ago. Other researchers believe that cooking was invented as late as 40,000 or 10,000 years ago. Evidence of fire is inconclusive, as wildfires started by lightning-strikes are still common in East Africa and other wild areas, and it is difficult to determine when fire was first used for cooking, as opposed to just being used for warmth or for keeping predators away.

Wrangham proposed cooking was instrumental in human evolution, as it reduced the time required for foraging and led to an increase in brain size. Since meat has a higher energy density than vegetables, and cooking it allows more nutrients to be liberated to the body, the introduction of cooked meat in the human diet reduced the energy requirements of the digestive system. He estimates the percentage decrease in gut size of early humans directly correlates to the increase in brain size. Most other anthropologists, however, oppose Wrangham, stating that archeological evidence suggests that cooking fires began in earnest only c.250,000 years ago, when ancient hearths, earth ovens, burnt animal bones, and flint appear across Europe and the Middle East. Two million years ago, the only sign of fire is burnt earth with human remains, which most other anthropologists consider to be mere coincidence rather than evidence of intentional fire. The mainstream view among anthropologists is that the increases in human brain-size occurred well before the advent of cooking, due to a shift away from the consumption of nuts and berries to the consumption of meat.

Food has become a part of material culture, and cuisine is much more than a substance. In the seventeenth and eighteenth centuries, food was a classic marker in Europe. However, in the nineteenth century, cuisine became a defining symbol of national identity. The discovery of the New World represented a major turning point in the history of food because of the movement of foods from and to Europe, such as potatoes, tomatoes, corn, yams, and beans. Food in America consisted of traditions that were adapted from England, but up until the end of this century, the presence of new ingredients along with the contact between diverse ethnic groups influenced experimentation. Industrialization was also a turning point that changed how food affected the nation.

During the period of industrialization, food began to be mass produced, mass marketed, and standardized. Factories processed, preserved, canned, and packaged a wide variety of foods, and processed cereals quickly became a defining feature of the American breakfast. In the twenties, freezing methods as well as the earliest cafeterias and fast food establishments emerged. This point in time is when processed and nationally distributed foods became a huge part of the nation's diet.

Along with changes in food, there have also been several changes in nutritional guidelines as well. Since 1916, there have been several different nutrition guidelines issued by the United States government, eventually leading up to the food pyramid. In 1916, "Food For Young Children" along with its sequel for adults, "How to Select Foods" was the first USDA guide to give specific dietary guidelines. Updated in the 1920s to these guides gave shopping suggestions for different-sized families along with a Depression Era revision which included four cost levels. In 1943, the USDA created the "Basic Seven" chart to make sure that people got the recommended nutrients. It included the first-ever Recommended Daily Allowances from the National Academy of Sciences. In 1956, the "Essentials of an Adequate Diet" brought recommendations which cut seven down to four groups that school children would learn about for decades. In 1979, a guide called "Food" was published, which addressed the link between too much of certain foods and chronic diseases. This publication also added "fats, oils, and sweets" to the four basic food groups and cautioned moderation. In 1992, the food pyramid was debuted. The USDA introduced this, which represented proportions of foods in a balanced diet. In 2005, the pyramid got a makeover and was renamed MyPyramid. Lastly, in 2011, the "Plate and the Moon" theory came about.

Ingredients in Cooking

Most ingredients in cooking are derived from living organisms. Vegetables, fruits, grains and nuts as well as herbs and spices come from plants, while meat, eggs, and dairy products come from animals. Mushrooms and the yeast used in baking are kinds of fungi. Cooks also use water and minerals such as salt. Cooks can also use wine or spirits. Naturally occurring ingredients contain various amounts of molecules called proteins, carbohydrates and fats. They also contain water and minerals. Cooking involves a manipulation of the chemical properties of these molecules.

Carbohydrates

Carbohydrates include the common sugar, sucrose (table sugar), a disaccharide, and such simple sugars as glucose (from the digestion of table sugar) and fructose (from fruit), and starches from sources such as cereal flour, rice, arrowroot, and potato. The interaction of heat and carbohydrate is complex.

Long-chain sugars such as starch tend to break down into simpler sugars when cooked, while simple sugars can form syrups. If sugars are heated so that all water of crystallisation is driven off, then caramelization starts, with the sugar undergoing thermal decomposition with the formation of carbon, and other breakdown products producing caramel. Similarly, the heating of sugars and proteins elicits the Maillard reaction, a basic flavor-enhancing technique.

An emulsion of starch with fat or water can, when gently heated, provide thickening to the dish being cooked. In European cooking, a mixture of butter and flour called a roux is used to thicken liquids to make stews or sauces. In Asian cooking, a similar effect is obtained from a mixture of rice or corn starch and water. These techniques rely on the properties of starches to create simpler mucilaginous saccharides during cooking, which causes the familiar thickening of sauces. This thickening will break down, however, under additional heat.

Fats

Types of fat include vegetable oils, animal products such as butter and lard, as well as fats from grains, including corn and flax oils. Fats can reach temperatures higher than the boiling point of water, and are often used to conduct high heat to other ingredients, such as in frying or sautéing.

Proteins

Edible animal material, including muscle, offal, milk, eggs and egg whites, contains substantial amounts of protein. Almost all vegetable matter (in particular legumes and seeds) also includes proteins, although generally in smaller amounts. Mushrooms have high protein content. Any of these may be sources of essential amino acids. When proteins are heated they become denatured (unfolded) and change texture. In many cases, this causes the structure of the material to become softer or more friable – meat becomes cooked and is more friable and less flexible. In some cases, proteins can form more rigid structures, such as the coagulation of albumen in egg whites. The formation of a relatively rigid but flexible matrix from egg white provides an important component in baking cakes, and also underpins many desserts based on meringue.

Vitamins and minerals

Vitamins are materials required for normal metabolism but which the body cannot manufacture itself and which must therefore come from soil. Vitamins come from a number of sources including fresh fruit and vegetables (Vitamin C), carrots, liver (Vitamin A), cereal bran, bread, liver e (B vitamins), fish liver oil (Vitamin D) and fresh green vegetables (Vitamin K). Many minerals are also essential in small quantities including iron, calcium, magnesium and sulphur; and in very small quantities copper, zinc and selenium. The micronutrients, minerals, and vitamins in fruit and vegetables may be destroyed or eluted by cooking. Vitamin C is especially prone to oxidation during cooking and may be completely destroyed by protracted cooking.

Water

Cooking often involves water, which is frequently present in other liquids, both added in order to immerse the substances being cooked (typically water, stock or wine), and released from the foods themselves. Liquids are so important to cooking that the name of the cooking method used is often based on how the liquid is combined with the food, as in steaming, simmering, boiling, braising, and blanching. Heating liquid in an open container results in rapidly increased evaporation, which concentrates the remaining flavor and ingredients – this is a critical component of both stewing and sauce making.

Cooking Techniques

There are very many methods of cooking, most of which have been known since antiquity. These include baking, roasting, frying, grilling, barbecuing, smoking, boiling, steaming and braising. A more recent innovation is microwaving. Various methods use differing levels of heat and moisture and vary in cooking time. The method chosen greatly affects the end result. Some foods are more appropriate to some methods than others. Some major hot cooking techniques include:

Roasting

Roasting is a cooking method that uses dry heat, whether an open flame, oven, or other heat source. Roasting can enhance flavor through caramelization and Maillard browning on the surface of the food. Roasting uses indirect, diffused heat (as in an oven), and is suitable for slower cooking of meat in a larger, whole piece. Meats and most root and bulb vegetables can be roasted. Any piece of meat, especially red meat, that has been cooked in this fashion is called a roast. In addition, large uncooked cuts of meat are referred to as roasts. A roast joint of meat can take one, two, even three hours to cook - the resulting meat is tender. Also, meats and vegetables prepared in this way are described as "roasted", e.g., roasted chicken or roasted squash.

Methods

For roasting, the food may be placed on a rack, in a roasting pan or, to ensure even application of heat, may be rotated on a spit or rotisserie. If a pan is used, the juice can be retained for use in gravy, Yorkshire pudding, etc. During oven roasting, hot air circulates around the meat, cooking all sides evenly. There are several plans for roasting meat: low-temperature cooking, high-temperature cooking, and a combination of both. Each method can be suitable, depending on the food and the tastes of the people.

A low-temperature oven, 95 °C to 160 °C (200 °F to 325 °F), is best when cooking with large cuts of meat, turkey and whole chickens. This is not technically roasting temperature, but it is called slow-roasting. The benefit of slow-roasting an item is less moisture loss and a more tender product. More of the collagen that makes meat tough is dissolved in slow cooking. At true roasting temperatures, 200 °C (400 °F) or more, the water inside the muscle is lost at a high rate.

Cooking at high temperatures is beneficial if the cut is tender enough—as in filet mignon or strip loin—to be finished cooking before the juices escape. A reason for high temperature roasting is to brown the outside of the food, similar to browning food in a pan before pot roasting or stewing it. Fast cooking gives more variety of flavor, because the outside is brown while the center is much less done.

The combination method uses high heat just at either the beginning or the end of the cooking process, with most of the cooking at a low temperature. This method produces the golden-brown texture and crust, but maintains more of the moisture than simply cooking at a high temperature, although the product will not be as moist as low-temperature cooking the whole time. Searing and then turning down to low is also beneficial when a dark crust and caramelized flavor is desired for the finished product. Note that searing in no way "locks in" moisture: moisture loss is simply a function of heat and time. The outside is brown and the rest is done fairly uniformly.In general, in either case, the meat is removed from heat before it has finished cooking and left to sit for a few minutes, while the inside cooks further from the residual heat content, a phenomenon known as carry over cooking, also known as "resting" the meat.

The objective in any case is to retain as much moisture as possible, while providing the texture and color. As meat cooks, the structure and especially the collagen breaks down, allowing juice to come out of the meat. So meat is juiciest at about medium rare while the juice is coming out. During roasting, meats and vegetables are frequently basted on the surface with butter, lard, or oil to reduce the loss of moisture by evaporation. In recent times, plastic oven bags have become popular for roasts. These cut cooking times and reduce the loss of moisture during roasting, but reduce flavor development from Maillard browning, somewhat more like (boiled or steamed) stew or pot roast. They are particularly popular for turkeys.

Until the late 19th century, roasting by dry heat in an oven was called baking. Roasting originally meant turning meat or a bird on a spit in front of a fire. It is one of the oldest forms of cooking known.

Traditionally recognized roasting methods consist only of baking and cooking over or near an open fire. Grilling is normally not technically a roast, since a grill (gridiron) is used. Barbecuing and Smoking differ from roasting because of the lower temperature and controlled smoke application.

Meat

Most meat roasts are large cuts of meat. Many roasts are tied with string prior to roasting, often using the reef knot or the packer's knot. Tying holds them together during roasting, keeping any stuffing inside, and keeps the roast in a round profile, which promotes even cooking.

Prior to roasting in an oven, meat is generally "browned" by brief exposure to high temperature. This imparts a traditional flavor and color to the roast. Red meats such as beef, lamb, and venison, and certain game birds are often roasted to be "pink" or "rare", meaning that the center of the roast is still red. Roasting is a preferred method of cooking for most poultry, and certain cuts of beef, pork, or lamb. Although there is a growing fashion in some restaurants to serve "rose pork", temperature monitoring of the center of the roast is the only sure way to avoid foodborne disease.

Vegetables

Some vegetables, such as potatoes, zucchini, pumpkin, turnips, parsnips, cauliflower, asparagus, squash, and peppers lend themselves to roasting as well. Roasted chestnuts are also a popular snack in winter.

Barbecuing

Barbecue is a method and apparatus for char grilling food in the hot smoke of a wood fire, usually charcoal fueled. In the United States, to grill is to cook in this manner quickly, while barbecue is typically a much slower method utilizing less heat than grilling, attended to over an extended period of several hours.

The term as a noun can refer to the meat, the cooking apparatus itself (the "barbecue grill" or simply "barbecue") or to the party that includes such food or such preparation methods. The term as an adjective can refer to foods cooked by this method. The term is also used as a verb for the act of cooking food in this manner.

Barbecue is usually done in an outdoor environment by cooking and smoking the meat over wood or charcoal. Restaurant barbecue may be cooked in large brick or metal ovens specially designed for that purpose.

Styles

In British usage, barbecuing refers to a fast cooking process directly over high heat, while grilling refers to cooking under a source of direct, high heat—known in the United States and Canada as broiling. In American English usage, however, grilling refers to a fast process over high heat, while barbecuing refers to a slow process using indirect heat and/or hot smoke (very similar to some forms of roasting). For example, in a typical U.S. home grill, food is cooked on a grate directly over hot charcoal, while in a U.S. barbecue, the coals are dispersed to the sides or at significant distance from the grate. Its South American versions are the southern Brazilian churrasco and the Argentine asado.

Alternatively, an apparatus called a smoker with a separate fire box may be used. Hot smoke is drawn past the meat by convection for very slow cooking. This is essentially how barbecue is cooked in most U.S. "barbecue" restaurants, but nevertheless, many consider this to be a distinct cooking process called hot smoking.

American South and Midwest

According to estimates, prior to the American Civil War, Southerners ate around five pounds of pork for every one pound of beef they consumed. Because of the poverty of the southern United States at this time, every part of the pig was eaten immediately or saved for later (including the ears, feet, and other organs). Because of the effort to capture and cook these wild hogs, pig slaughtering became a time for celebration, and the neighborhood would be invited to share in the largesse. In Cajun culture, these are called boucheries. These feasts are sometimes called 'pig pickin's.' The traditional Southern barbecue grew out of these gatherings.

Each Southern locale has its own particular variety of barbecue, particularly concerning the sauce. North Carolina sauces vary by region; eastern North Carolina uses a vinegar-based sauce, the center of the state enjoys Lexington-style barbecue, which uses a combination of ketchup and vinegar as their base, and western North Carolina uses a heavier ketchup base. Lexington boasts of being "The Barbecue Capital of the World" and it has more than one BBQ restaurant per 1,000 residents. South Carolina is the only state that includes all four recognized barbecue sauces, including mustard-based, vinegar-based, and light and heavy tomato-based. Memphis barbecue is best known for tomato- and vinegar-based sauces. In some Memphis establishments and in Kentucky, meat is rubbed with dry seasoning (dry rubs) and smoked over hickory wood without sauce; the finished barbecue is then served with barbecue sauce on the side.

The barbecue of Alabama, Georgia, and Tennessee is almost always pork served with a sweet tomato-based sauce. However, several regional variations exist as well. Alabama is particularly known for its distinctive white sauce, a mayonnaise- and vinegar-based sauce, originating in northern Alabama, used predominantly on chicken and pork. A popular item in North Carolina and Memphis is the pulled pork sandwich served on a bun and often topped with coleslaw. Pulled pork is prepared by shredding the pork after it has been barbecued.

Kansas City-style barbecue is characterized by its use of different types of meat (including pulled pork, pork ribs, burnt ends, smoked sausage, beef brisket, beef ribs, smoked/grilled chicken, smoked turkey, and sometimes fish), a variety attributable to Kansas City's history as a center for meat packing in the U.S. Hickory is the primary wood used for smoking in KC, while the sauces are typically tomato based with sweet, spicy and tangy flavor profiles. Burnt ends, the flavorful pieces of meat cut from the ends of a smoked beef or pork brisket, are popular in many Kansas City-area barbecue restaurants.

Pit-beef prevails in Maryland and is often enjoyed at large outdoor "bull roasts", which are common for club or association fundraising events. Maryland-style pit-beef is not the product of barbecue cookery in the strictest sense, as there is no smoking of the meat involved; rather, it involves grilling the meat over a high heat. The meat is typically served rare, with a strong horseradish sauce as the preferred condiment.

The state of Kentucky, particularly Western Kentucky, is unusual in its barbecue cooking, in that the preferred meat is mutton. This kind of mutton barbecue is often used in communal events in Kentucky, such as political rallies, county fairs and church fund-raising events.

In much of the world outside of the American South, barbecue has a close association with Texas. Many barbecue restaurants outside the United States claim to serve "Texas barbecue", regardless of the style they actually serve. Texas barbecue is often assumed to be primarily beef. This assumption, along with the inclusive term "Texas barbecue", is an oversimplification. Texas has four main styles, all with different flavors, different cooking methods, different ingredients, and different cultural origins. In the June 2008 issue of Texas Monthly Magazine Snow's BBQ in Lexington was rated as the best BBQ in the state of Texas. This ranking was reinforced when New Yorker Magazine also claimed that Snow's BBQ was "The Best Texas BBQ in the World".

China

In China, as early as the protohistory, human use fire to cook meat, so that it may be the most original barbecue culture and it can be traced back to the Fuxi era. We all know a legend called Fuxi Fuxi take a fire, then take the fish, birds on the fire, and told people to eat. People eat tasty barbecue, then their body will be healthier. Then people call Fuxi as the first person who cook meats with fire.

At the beginning, people only grill meat such as fish, beef, and pork. As time goes on, people use vegetable and mushrooms as ingredients. Besides, people use several spices such as pepper and cloves to flavor ingredients.

Korea

Genghis Khan Teppanyaki grill is from Temujin era, Genghis Khan had always fought for the territory. After the war, Genghis Khan, ordered soldiers using helmet to grill. This barbecue is named "Genghis Khan Barbecue" by Chinese, and is named " Genghis Khan Teppanyaki by Japanese. The original Genghis Khan Teppanyaki ingredient grill is lamb chops. However, nowadays, people not only grill lamb but other meats such as fish, chicken, beef and so on.

As the descendants of Genghis Khan conquer Koryo (a place in Korean), Genghis Khan Teppanyaki barbecue was introduced to the Korean peninsula. Although Genghis Khan's grandson Kublai failed to conquer Japan, Genghis Khan Teppanyaki barbecue not only spread to Japan but on the Korean Peninsula.

The original device of Korean barbecue is iron plate which is designed as similar to a stair to drain away the excess oil. And, there is a box at the bottom of the plate to collect the excess oil. Then, the Korean wraps meats with vegetable such as lettuces. Unlike western, Korean likes wrapping vegetable with meats as they like lite food. Spicy food does agree with Korean because it is very cold in winter.

Japan

Japanese especially people in Hokkaido still likes Genghis Khan Teppanyaki until now. They improve the iron plate to an easier way to grill but it still like a helmet.

Events and Gatherings

The word barbecue is also used to refer to a social gathering where food is served, usually outdoors in the late afternoon or evening. In the southern United States, outdoor gatherings are not typically called "barbecues" unless barbecue itself will actually be on the menu, instead generally favoring the word "cookouts". The device used for cooking at a barbecue is commonly referred to as a "barbecue", "barbecue grill", or "grill". In North Carolina, however, "barbecue" is a noun primarily referring to the food and never used by native North Carolinians to describe the act of cooking or the device on which the meat is cooked.

Often referred to as "The World Series of Barbecue", The American Royal Barbecue Contest is held each October in Kansas City, Missouri. This event comprises two distinct competitions held over the course of four days. The first contest is the Invitational Contest, with competing teams being required to obtain an invitation by winning other qualifying contests throughout the year. The second competition is an open contest that any team can compete in. This open contest is the largest championship barbecue competition in the world, with the 2007 event attracting 496 teams.

The World Championship Barbecue Cooking Contest is held annually in Memphis, Tennessee, during the Memphis in May festival. According to the Guinness Book of World Records, it is the world's largest pork barbecue contest.

Other barbecue competitions are held in virtually every state in the United States during the warmer months, usually beginning in April and going through September. These events feature keen competitions between teams of cooks and are divided into separate competitions for the best pork, beef and poultry barbecue and for the best barbecue sauces.

Techniques

Barbecuing encompasses four or five distinct types of cooking techniques. The original technique is cooking using smoke at lower temperatures (usually around 240–270 °F or 115–125 °C) and significantly longer cooking times (several hours), known as smoking. Another technique is baking, utilizing a masonry oven or any other type of baking oven, which uses convection to cook meats and starches with moderate temperatures for an average cooking time (about an hour plus a few extra minutes). Yet another technique is braising, which combines direct dry heat charbroiling on a ribbed surface with a broth-filled pot for moist heat, cooking at various speeds throughout the duration (starting fast, slowing down, then speeding up again, lasting for a few hours). Finally, grilling is done over direct dry heat, usually over a hot fire (i.e., over 500 °F (260 °C)) for a short time (minutes). Grilling may be done over wood, charcoal, gas (natural gas or propane), or electricity.

Grilling

Grilling is a form of cooking that involves dry heat applied to the surface of food, commonly from above or below. It is sometimes referred to as barbecuing but that word can also mean a different cooking technique.

Grilling usually involves a significant amount of direct, radiant heat, and tends to be used for cooking meat quickly and meat that has already been sliced (or other pieces). Food to be grilled is cooked on a grill (an open wire grid such as a gridiron with a heat source above or below), a grill pan (similar to a frying pan, but with raised ridges to mimic the wires of an open grill), or griddle (a flat plate heated from below). Heat transfer to the food when using a grill is primarily via thermal radiation. Heat transfer when using a grill pan or griddle is by direct conduction. In the United States and Canada, when the heat source for grilling comes from above, grilling is termed broiling. In this case, the pan that holds the food is called a broiler pan, and heat transfer is by thermal radiation.

Direct heat grilling can expose food to temperatures often in excess of 260 °C (500 °F). Grilled meat acquires a distinctive roast aroma from a chemical process called the Maillard reaction. The Maillard reaction only occurs when foods reach temperatures in excess of 155 °C (310 °F).

Studies have shown that cooking beef, pork, poultry, and fish at high temperatures can lead to the formation of heterocyclic amines, benzopyrenes, and polycyclic aromatic hydrocarbons, which are carcinogens. Marination may reduce the formation of these compounds. Grilling is often presented as a healthy alternative to cooking with oil, although the fat and juices lost by grilling can contribute to drier food.

Regional Variations

Asia

In Japanese cities, a yakitori cart, restaurant, or shop with charcoal-fired grills and marinated grilled meat on a stick can often be found. Yakiniku, is a type of food where meat and/or vegetables are grilled directly over small charcoal or gas grills at high temperatures.In Malaysia, Singapore, Indonesia, and Thailand, a popular food item from food vendors is satay, marinated meat on a bamboo skewer grilled over a charcoal fire and served with peanut (sate) sauce.

Germany

In Germany, the most prominent outdoor form of grilling is by using charcoal in a kettle. Besides charcoal, sometimes gas and electric heat sources are used, too. Other methods are used less frequently.

South America

In Argentina and Uruguay, both asado (beef roasted on a fire) and steak a la parrilla (beefsteak cooked on traditional grill) are staple dishes and even hailed as national specialities.

UK, Commonwealth and Ireland

In the United Kingdom, Commonwealth countries and Ireland, grilling generally refers to cooking food directly under a source of direct, dry heat. The "grill" is usually a separate part of an oven where the food is inserted just under the element. This practice is referred to as "broiling" in North America.In Australia, grilling can refer to cooking food directly under a source of direct heat or with heat from below, as in the US.

In electric ovens, grilling may be accomplished by placing the food near the upper heating element, with the lower heating element off and the oven door partially open. Grilling in an electric oven may create a large amount of smoke and cause splattering in the oven. Gas ovens often have a separate compartment for grilling, such as a drawer below the flame.

United States

In the United States, the use of the word grill refers to cooking food directly over a source of dry heat, typically with the food sitting on a metal grate that leaves "grill marks." Grilling is usually done outdoors on charcoal grills or gas grills, a recent trend is the concept of infrared grilling. Grilling may also be performed using stove-top "grill pans" which have raised metal ridges for the food to sit on, or using an indoor electric grill.

A skewer or brochette, or a rotisserie may be used to cook small pieces of food. The resulting food product is often called a "kabob" or "kebab" which means "to grill" in Persian, which is short for "shish kebab" (shish = skewer)(similar to a "satay" in Asian cuisine, or "alambre" in Mexican-Yucatán cuisine). Shish kebabs have a Persian origin, but are now commonplace in American cuisine.

Mesquite or hickory wood chips (damp) may be added on top of the coals to allow a smoldering effect that provides additional flavor to the food. Other hardwoods such as pecan, apple, maple and oak may also be used.

Risks of Grilling

As is true of any high-temperature frying or baking, when meat is grilled at high temperatures, the cooking process can generate carcinogenic chemicals.) Two processes are thought to be responsible. Heterocyclic amines - "HCAs are formed when amino acids, sugars, and creatine react at high temperatures." Additionally, polycyclic aromatic hydrocarbons - "PAHs are formed when fat and juices from meat grilled directly over an open fire drip onto the fire, causing flames. These flames contain PAHs that then adhere to the surface of the meat. PAHs can also be formed during other food preparation processes, such as smoking of meats."

However it is possible to significantly reduce carcinogens when grilling meat, or mitigate their effect. * Garlic, rosemary, basil, mint, sage, savory, marjoram, oregano, olive oil, cherries, and vitamin E have been shown to reduce formation of both HCAs and PCAs. Another method is pre-cooking the meat in the microwave, then draining meat juices so they do not fall onto flames, preventing release of PCAs. Side dishes and drinks rich in antioxidants, such as tea, have also been shown to neutralize the toxins by mixing in one's stomach

Grilling Methods

Gridironing

Gridironing is the cooking of meats or other foods using a grill suspended above a heat source. Grilling is often performed outdoors, using charcoal (real wood or preformed briquettes), wood, or propane gas. Food is cooked using direct radiant heat. Some outdoor grills include a cover so they can be used as smokers or for grill-roasting/barbecue. The suspended metal grate is often referred to as a gridiron.

Outdoor grilling on a gridiron may be referred to as "barbecue", though in US usage, the term barbecue refers to the cooking of meat by indirect heat and smoke. Barbecue has several meanings and may also be used to refer to the grilled food itself, to a distinct type of cooked meat called Southern barbecue, to the grilling device used to cook the food (a barbecue grill), or to the social event of cooking and eating such food (which may also be called a cook-out or braai).

Charcoal kettle-grilling

Charcoal kettle-grilling refers to the process of grilling over a charcoal fire in a kettle, to the point that the edges are charred, or charred grill marks are visible. Some restaurants seek to re-create the charcoal-grilled experience via the use of ceramic lava rocks or infrared heat sources, offering meats that are cooked in this manner as "charcoal-cooked" or "charcoal-grilled".

Grill-baking

By using a baking sheet pan placed above the grill surface, as well as a drip pan below the surface, it is possible to combine grilling and roasting to cook meats that are stuffed or coated with breadcrumbs or batter, as well as to bake breads and even casseroles and desserts. When cooking stuffed or coated meats, the foods can be baked first on the sheet pan, and then placed directly on the grilling surface for char marks, effectively cooking twice; the drip pan will be used to capture any crumbs that fall off from the coating or stuffing.

Grill-braising

It is possible to braise meats and vegetables in a pot on top of a grill. A gas or electric grill would be the best choices for what is known as "barbecue-braising" or "grill-braising", or combining grilling directly on the surface and braising in a pot. To braise on a grill, put a pot on top of the grill, cover it, and let it simmer for a few hours. There are two advantages to barbecue-braising: the first is that this method now allows for browning the meat directly on the grill before the braising, and the second is that it also allows for glazing the meat with sauce and finishing it directly over the fire after the braising, effectively cooking the meat three times, which results in a soft textured product that falls right off the bone. This method of cooking is slower than regular grilling but faster than pit-smoking, starting out fast, slowing down, and then speeding up again to finish; if a pressure cooker is used, the cooking time will be much faster.

Indoor grilling

Many restaurants incorporate an indoor grill as part of their cooking apparatus. These grills resemble outdoor grills, in that they are made up of a grid suspended over a heat source. Indoor grills are more likely to use electric or gas-based heating elements, however. Some manufacturers of residential cooking appliances now offer indoor grills for home use, either incorporated into a stovetop or as standalone electric devices.

Sear grilling

Sear-grill and gear grilling is a process of searing meat or food items with an infrared grill. In sear grilling, propane or natural gas is used to heat a ceramic plate, which then radiates heat at temperatures over 480°C.

Sear-grilling instantly sears the outside of meat to make the food more flavorful. Commonly, grilling heats the surrounding air to cook food. Instead, the infrared grill directly heats the food, not the air.

Stove-top pan grilling

Stove-top pan grilling is an indoor cooking process that uses a grill pan - a cooking pan similar to a frying pan but with raised ridges to emulate the function or look of a gridiron. In pan grilling, heat is applied directly to the food by the raised ridges, and also indirectly by heat radiating off the lower pan surface via the stove-top flame. Stove-top grill pans can also be used to put sear marks on meat before it is finished via overhead radiant heat. When cooking leaner meats, oil is often applied to the pan ridges to aid in food release.

Some griddles designed for stove-top use also incorporate raised ridges in addition to a flat cooking area. These are either on half of the cooking surface, or, in the case of reversible two-sided griddles, on one side with the flat surface on the other.

Flattop grilling

Foods termed "grilled" may actually be prepared on a hot griddle, or flat pan. The griddle or pan may be prepared with oil (or butter), and the food is cooked quickly over a high heat. Griddle-grilling is best for relatively greasy foods such as sausages. Some griddle-grilled foods may have grill marks applied to them during the cooking process with a branding plate, to mimic the appearance of charbroil-cooked food.

A flattop grill is a cooking appliance that resembles a griddle but performs differently because the heating element is circular rather than straight (side to side). This heating technology creates an extremely hot and even cooking surface, as heat spreads in a radial fashion over the surface.

The first flattop grills originated in Spain and are known as planchas or la plancha. Food that is cooked a la plancha means grilled on a metal plate. Plancha griddles or flat tops are chrome plated which prevents reaction with the food. Some base metal griddles will impart a subtle flavor to the food being cooked.

The flattop grill is a versatile platform for many cooking techniques such as sautéing, toasting, steaming, stir frying, grilling, baking, braising, and roasting, and can also be used in flambéing. In addition, pots and pans can be placed directly on the cooking surface for even more cooking flexibility. In most cases, the steel cooking surface is seasoned like cast iron cookware, providing a natural non-stick surface.

Charbroiling

Charbroiling, or chargrilling outside North America, refers to grilling on a surface with wide raised ridges, to the point of having the food slightly charred in texture. The phrase "put it in the broiler" is translated as "put it over/under the grill."

Overhead grilling

In the United States, oven pan broiling refers to a method of cooking on a broil pan with raised ridges, inside an oven, when the heat can be applied from either above or below. In gas and electric ovens, this is accomplished with a heating element and a broil pan. Sometimes, the food is placed near the upper heating element to intensify the heat. The lower heating element may or may not be left off and the oven door is sometimes opened partially. Gas ovens often have a separate compartment for broiling, sometimes a drawer below the bottom flame.

Salamander

A salamander is a culinary broiler characterized by very high temperature overhead infrared heating elements which may be powered by either electricity or gas. It is used primarily in professional kitchens for overhead grilling. It is also used for toasting, as well as browning of gratin dishes, melting cheeses onto sandwiches, and caramelising desserts such as crème brûlée.

Salamanders are generally similar to an oven without a front door, with the heating elements at the top. They are more compact: typically only half the height and depth of a conventional oven. They are often wall mounted at eye level enabling easy access and close control of the cooking process. Many salamanders can be fitted with a cast iron "branding" plate which is used to make grill marks on the surface of meat. Some grills can also be fitted with a rotisserie accessory for roasting meats.

Overhead heat has the advantage of allowing foods containing fats, such as steaks, chops and other cuts of meat, to be grilled without the risk of flare-ups caused by the rendered fat dripping into the heat source. The salamander's facility for extremely high temperature also takes less cooking time than other grilling techniques, reducing preparation times, which is a benefit in professional kitchens during a busy meal service.

Modern electric or gas salamanders take their name from the earlier salamander, an iron disc on a handle which is heated and placed over a dish to brown it, which in turn is named after the legendary salamander, an amphibian that was mythically believed to be immune to fire. In the 18th century, a salamander was the tool of choice for toasting the top of a dish. It consisted of a thick plate of iron attached to the end of a long handle with 2 feet, or rests, arranged near the end (where the iron plate is) for propping the plate over the food to be browned.

Two-sided grilling

Some commercial devices permit the simultaneous grilling of both sides of the meat at the same time.

The flame-grilling machine at Burger King, Carl's Jr./Hardee's, and other restaurants is called a 'broiler'. It works by moving meat patties along a chain conveyor belt between top and bottom burners, grilling both sides of the meat patty at the same time. This concept was invented in 1898, when the Bridge and Beach Co. of St. Louis, Missouri, started manufacturing a vertical cast iron stove. These stoves were designed to allow the meat to be flame-broiled (flame-grilled) on both sides at the same time. Custom hinged steel wire gridirons were built for use in the vertical broilers. The hinged gridirons were slid in and out of the stoves holding the meat while it cooked evenly on both sides, like modern day oven racks. These stoves took up a small amount of counter space. They were used in lunch spots to feed factory workers. One famous example of a vertical grill still in use is the purported inventor of the hamburger, Louis' Lunch, in New Haven, CT.

During the 1990s, double-sided grilling was popular in the USA using consumer electrical grills. US marketers of electric double-sided grilling appliances opted for the global term 'grilling' rather than the geographically isolated term "broiler." Hinged double-sided grills are generically known as contact grills.

Stone grills

Sometimes a stone is used to grill foods. Stones in these cases can store temperatures up to 450 °C (842 °F). Foods grilled on stone involve no fats or oil and are considered a healthier alternative.

Rotisserie

Rotisserie is a style of roasting where meat is skewered on a spit - a long solid rod used to hold food while it is being cooked over a fire in a fireplace or over a campfire, or roasted in an oven. This method is generally used for cooking large joints of meat or entire animals, such as pigs, turkeys, goats or historically, entire cattle. The rotation cooks the meat evenly in its own juices and allows easy access for continuous self-basting.

In medieval and early modern kitchens, the spit was the preferred way of cooking meat in a large household. A servant, preferably a boy, sat near the spit turning the metal rod slowly and cooking the food; he was known as the "spit boy" or "spit jack". Mechanical turnspits ("roasting jacks") were later invented, first powered by dogs on treadmills, and then by steam power and mechanical clockwork mechanisms. The spit could also be powered by a turbine mounted in the chimney with a worm transmission for torque and speed conversion. Spits are now usually driven by electric motors.

Rotisserie can also refer to a mechanical device used for rotisserie cooking, or to a restaurant specializing in spit-roasted meat. The word comes from French where it first appeared in Paris shops around 1450. Additionally, in restaurants employing the Escoffierian brigade de cuisine, the rotisseur is the chef responsible for all spit-roasted, oven roasted, grilled and in some cases fried foods.

Horizontal Rotisserie

This style of rotisserie mounts the spit horizontally. They are often used to cook whole chickens or roasts of various meats including beef and pork. The design may include a single spit mounted over an open broiler or grill, a single spit mounted within an otherwise-conventional oven, or many spits mounted within a large industrial oven. The latter are commonly used to mass produce roasted meats for sale to consumers.

In this style of rotisserie, balance is important. If the object to be cooked is far out of balance, it will impose a heavy load on the drive mechanism or cause the mechanism to fail to turn. Loose chicken legs or wings can also cause the mechanism to jam. For these two reasons, some skewering skill is required.

High-end consumer ovens commonly come with a rotisserie (or allow the installation of a rotisserie as an option). In these cases, the motor drive mechanism is usually concealed within the oven. The rotisserie is used by removing the normal cooking racks; a special carrier may be needed to provide one or both bearing points for the spit.

The speed of rotation can vary, depending on the type of food being cooked and proximity to the heat source. Large animals, such as whole pigs, are often cooked at speeds ranging from 3 rpm to 1 rph. Ron Popeil, inventor of the Ronco "Showtime" rotisserie oven, claims that 6 rpm is an ideal speed for cooking a variety of food. Generally speaking, when a whole animal is being spit roasted, it is best rotated faster in the early part of the roasting, to help interior temperature rise without burning the exterior. Once the exterior has started to crust, then the rate of rotation is adjusted down so as to minimize dripping. This allows juices and drippings to stay on, thus maximizing flavor. The latter is particularly important if using a horizontal spit and coals or an open fire, in which case the drippings cannot be salvaged for basting. For a horizontal spit it is also preferable to keep the roast further away from the heat source in the beginning, which minimizes infrared heating while preserving air convection heating. Later on, the roast is brought closer, thus combining intense convection and radiation heating.

Supermarkets commonly mass-produce rotisserie chickens using a vertical rotisserie that has metal bars to hold the chicken in place through the weakest part of the breast (also which hardly affects the meat itself from the impaling but still holds the chicken firmly in place) and the densest part of the chicken located just below the drumstick of the chicken.

Vertical Rotisserie

The other common style of rotisserie is the vertical rotisserie in which the heat is applied directly from the side or, less-commonly, convected up from below. In this style of rotisserie, balance of the load is less important than with a horizontal rotisserie.

Searing

Searing (or pan searing) is a technique used in grilling, baking, braising, roasting, sautéing, etc., in which the surface of the food (usually meat, poultry or fish) is cooked at high temperature so a caramelized crust forms. Similar techniques, browning and blackening, are typically used to sear all sides of a particular piece of meat, fish, poultry, etc. before finishing it in the oven. To obtain the desired brown or black crust, the meat surface must exceed 150 °C (300 °F), so searing requires the meat surface be free of water, which boils at around 100 °C (212 °F).

Although often said to "lock in the moisture" or "seal in the juices", searing has been demonstrated to result in a greater net loss of moisture versus cooking to the same internal temperature without first searing. Nonetheless, it remains an essential technique in cooking meat for several reasons:

— The browning creates desirable flavors through caramelization and the Maillard reaction.
— The appearance of the food is usually improved with a well-browned crust.
— The contrast in taste and texture between the crust and the interior makes the food more interesting to the palate.

Typically in grilling, the food will be seared over very high heat and then moved to a lower-temperature area of the grill to finish cooking. In braising, the seared surface acts to flavor, color and otherwise enrich the liquid in which the food is being cooked.

Sealing in the Juices

The belief that searing meat "seals in the juices" is widespread and still often repeated. This theory was first put forth by Justus von Liebig, a German chemist and food scientist, around 1850. The notion was embraced by contemporary cooks and authors, including Auguste Escoffier. It is more typically cited in regards to larger cuts, especially steaks and chops, of non-poultry meats such as beef, pork, lamb and tuna.

Simple experimentation can test the theory: cook two similar cuts of meat, searing one first and not the other. Then cook each normally in a preferred method (roasting, baking, grilling etc.) until each reaches the same exact predetermined internal temperature. Weigh the end results to see which loses more moisture. (The Food Network program Good Eats carried out such a test in episode EA1H22, Myth Smashers.) As early as the 1930s, such experiments were carried out; the seared roasts lost the same amount of moisture or more. (Generally more liquid is lost, since searing exposes the meat to higher temperatures that destroy more cells, in turn releasing more liquid.)

Moisture in liquid and vapor form continues to escape from a seared piece of meat. For this reason, searing is sometimes done at the end of the cooking process to gain the flavor benefits of the Maillard reaction, as well as the benefits of cooking for a greater duration with more wetness.

Baking

Baking is a food cooking method using prolonged dry heat acting by convection, and not by thermal radiation, normally in an oven, but also in hot ashes, or on hot stones. When the desired temperature is reached within the heating instrument, the food is placed inside and baked for a certain amount of time. It is primarily used for the preparation of bread, cakes, pastries and pies, tarts, quiches, cookies and crackers. Such items are sometimes referred to as "baked goods," and are sold at a bakery. A person who prepares baked goods as a profession is called a baker. It is also used for the preparation of baked potatoes, baked apples, baked beans, some casseroles and pasta dishes such as lasagne, and various other foods, such as the pretzel.

Many commercial ovens are provided with two heating elements: one for baking, using convection and thermal conduction to heat the food, and one for broiling or grilling, heating mainly by radiation. Meat may also be baked, but this is usually reserved for meatloaf, smaller cuts of whole meats, and whole meats that contain stuffing or coating such as bread crumbs or buttermilk batter; larger cuts prepared without stuffing or coating are more often roasted, a similar process, using higher temperatures and shorter cooking times. Baking can sometimes be combined with grilling to produce a hybrid barbecue variant, by using both methods simultaneously or one before the other, cooking twice. Baking is connected to barbecuing because the concept of the masonry oven is similar to that of a smoke pit.

The baking process does not require any fat to be used to cook in an oven. Some makers of snacks such as potato chips or crisps have produced baked versions of their snack items as an alternative to the usual cooking method of deep-frying in an attempt to reduce the calorie or fat content of their snack products.Baking has opened up doors to businesses like cake making factories and privatised cake shops but the baking process is done in a more larger mass in bigger and open furnaces.

The dry heat of baking changes the form of starches in the food and causes its outer surfaces to brown, giving it an attractive appearance and taste. The browning is caused by caramelization of sugars and the Maillard reaction. The moisture is never entirely "sealed in"; over time, an item being baked will become dry. This is often an advantage, especially in situations where drying is the desired outcome, like drying herbs or roasting certain types of vegetables. The most common baked item is bread. Variations in the ovens, ingredients and recipes used in the baking of bread result in the wide variety of breads produced around the world.

Some foods are surrounded with moisture during baking by placing a small amount of liquid (such as water or broth) in the bottom of a closed pan, and letting it steam up around the food, a method commonly known as braising or slow baking.

When baking, consideration must be given to the amount of fat that is contained in the food item. Higher levels of fat such as margarine, butter or vegetable shortening will cause an item to spread out during the baking process.

With the passage of time breads harden; they become stale. This is not primarily due to moisture being lost from the baked products, but more a reorganization of the way in which the water and starch are associated over time. This process is similar to recrystallization, and is promoted by storage at cool temperatures, such as in a domestic refrigerator.

In ancient history, the first evidence of baking occurred when humans took wild grass grains, soaked them in water, and mixed everything together, mashing it into a kind of broth-like paste. The paste was cooked by pouring it onto a flat, hot rock, resulting in a bread-like substance. Later, this paste was roasted on hot embers, which made bread-making easier, as it could now be made anytime fire was created.

Baking flourished in the Roman Empire. In about 300 BC, the pastry cook became an occupation for Romans (known as the pastillarium). This became a respected profession because pastries were considered decadent, and Romans loved festivity and celebration. Thus, pastries were often cooked especially for large banquets, and any pastry cook who could invent new types of tasty treats was highly prized. Around 1 AD, there were more than three hundred pastry chefs in Rome, and Cato wrote about how they created all sorts of diverse foods, and flourished because of those foods. Cato speaks of an enormous amount of breads; included amongst these are the libum (sacrificial cakes made with flour), placenta (groats and cress), spira (our modern day flour pretzels), scibilata (tortes), savaillum (sweet cake), and globus apherica (fritters). A great selection of these, with many different variations, different ingredients, and varied patterns, were often found at banquets and dining halls. The Romans baked bread in an oven with its own chimney, and had mills to grind grain into flour.

Eventually, because of Rome, the art of baking became known throughout Europe, and eventually spread to the eastern parts of Asia. Bakers often baked goods at home and then sold them in the streets. This scene was so common that Rembrandt illustrated a work that depicted a pastry chef selling pancakes in the streets of Germany, with children clamoring for a sample. In London, pastry chefs sold their goods from handcarts. This developed into a system of delivery of baked goods to households, and demand increased greatly as a result. In Paris, the first open-air café of baked goods was developed, and baking became an established art throughout the entire world.

Blind-baking

Baking blind (sometimes called pre-baking) is the process of baking a pie crust or other pastry without the filling. Blind baking a pie crust is necessary when it will be filled with an unbaked filling (such as with pudding or cream pies) or when the filling has a shorter bake time than the crust. Blind baking a pie crust also helps prevent the pie crust from becoming soggy from its filling.

Generally, the pie crust is lined with aluminium foil or parchment paper, then filled with pastry- or pie-weights to keep the crust's shape while baking. These weights are typically ceramic or metal beads, but dried peas, lentils, beans or other pulses can be used as baking beans as well. The crust is poked repeatedly with the tines of a fork to produce small holes—this helps the steam escape and prevents the crust from bubbling up. After the pie crust is done, the temporary filling is replaced with the proper filling.

Blind-baking also helps to form a nice pastry case for the filling as it has already been partially formed from blind-baking.

Boiling in Cooking

In cooking, boiling is the method of cooking food in boiling water, or other water-based liquid such as stock or milk. Simmering is gentle boiling, while in poaching the cooking liquid moves but scarcely bubbles.

Boiling is a very harsh technique of cooking. Delicate foods such as fish cannot be cooked in this fashion because the bubbles can damage the food. Foods such as red meat, chicken, and root vegetables can be cooked with this technique because of their tough texture.

The boiling point of water is typically considered to be 100 °C or 212 °F. Pressure and a change in composition of the liquid may alter the boiling point of the liquid. For this reason, high elevation cooking generally takes longer since boiling point is a function of atmospheric pressure. In Denver, Colorado, USA, which is at an elevation of about one mile, water boils at approximately 95 °C or 203 °F. Depending on the type of food and the elevation, the boiling water may not be hot enough to cook the food properly. Similarly, increasing the pressure as in a pressure cooker raises the temperature of the contents above the open air boiling point.

Adding a water soluble substance, such as salt or sugar also increases the boiling point. This is called boiling-point elevation. At palatable concentrations of salt, the effect is very small, and the boiling point elevation is difficult to notice. However, while making thick sugar syrup, such as for Gulab Jamun, one will notice boiling point elevation. Due to variations in composition and pressure, the boiling point of water is almost never exactly 100 °C, but rather close enough for cooking.

Bringing water to a boil is generally done by applying maximal heat, then shutting off when the water has come to a boil, which is known as bang–bang control. Keeping water at or below a boil requires more careful control of temperature, particularly by using feedback.

In places where the available water supply is contaminated with disease-causing bacteria, boiling water and allowing it to cool before drinking it is practiced as a valuable health measure. Boiling is the most certain way of killing all microorganisms in emergency situations.

Foods suitable for boiling include vegetables, starchy foods such as rice, noodles and potatoes, eggs, meats, sauces, stocks and soups.

Boiling has several advantages. It is safe and simple, and it is appropriate for large-scale cookery. Older, tougher, cheaper cuts of meat and poultry can be made digestible. Nutritious, well flavoured stock is produced. Also, maximum color and nutritive value is retained when cooking green vegetables, provided boiling time is kept to the minimum.

On the other hand, there are several disadvantages. There is a loss of soluble vitamins from foods to the water (if the water is discarded), and some boiled foods can look unattractive. Boiling can also be a slow method of cooking food.

Boiling can be done in several ways: The food can be placed into already rapidly boiling water and left to cook, the heat can be turned down and the food can be simmered; or the food can also be placed into the pot, and cold water may be added to the pot. This may then be boiled until the food is satisfactory.

Water on the outside of a pot, i.e. a wet pot, increases the time it takes the pot of water to boil. The pot will heat at a normal rate once all excess water on the outside of the pot evaporates.

Boiling is also often used to remove salt from certain foodstuffs, such as bacon, if a less saline product is required.

Blanching

Blanching is a cooking process wherein the food substance, usually a vegetable or fruit, is plunged into boiling water, removed after a brief, timed interval, and finally plunged into iced water or placed under cold running water (shocked) to halt the cooking process.

The meaning of blanching is “to whiten”, but this is not always the purpose of blanching in cooking. Food is blanched to soften it, or to partly or fully cook it, or to remove a strong taste (for example of bacon, cabbage, or onions).

When almonds or pistachios are blanched, the skin of the nut (botanically the seed coat surrounding the embryo) softens and can be easily removed later.

The technique of blanching vegetables is best illustrated by the requirement to stop cook asparagus, otherwise it would become soggy due to the asparagus continuing to cook after it is removed from the boiling water. To cook asparagus using the blanching method, the shoots are boiled for 30 seconds, and then to stop further cooking they are dipped into cold or ice water. Note: the term “blanched asparagus” is also used to refer to white asparagus, which is produced by a special gardening technique.

Blanching may simply mean boiling in water for an extended period to remove unpleasant flavours such as tannins. For example, a recommended treatment for African yam bean (Sphenostylis stenocarpa), is dehulling followed by blanching in water for 40 minutes at 100 °C.

It is recommended that protein-rich salads to be eaten cold (e.g., tuna, turkey, ham, shrimp, lobster, and chicken) are prepared by immersing these protein foods in boiling water for 30 seconds followed by fast chilling. In addition celery, "which is almost always a component of these salads ... should be treated so as to minimize its bacterial content".

In the case of French fries, blanching often refers to pre-cooking the French fries in oil at a lower temperature prior to finishing them at a higher temperature. The advantage is that the blanching step cooks the potato. The second step at the higher temperature crisps the outside.

In the early 1960s, McDonald's Restaurant in Charlotte, North Carolina, used a process on their potatoes they called 'blanching.' After the potatoes were mechanically peeled and then manually pushed through a slicer producing raw French fries, they were batch-soaked in room-temperature water for 3–5 minutes to remove starch, which they called blanching.

Braising

Braising is a combination cooking method using both moist and dry heat; typically the food is first seared at a high temperature and then finished in a covered pot with a variable amount of liquid, resulting in a particular flavor. Braising of meat is often referred to as pot roasting, though some authors make a distinction between the two methods based on whether additional liquid is added.

Braising relies on heat, time, and moisture to break down the tough connective tissue collagen in meat, making it an ideal way to cook tougher cuts. Many classic braised dishes such as coq au vin are highly evolved methods of cooking tough and otherwise unpalatable foods. Pressure cooking and slow cooking (e.g., crockpots) are forms of braising.

Most braises follow the same basic steps. The food to be braised (meat, poultry, but also vegetables or mushrooms) is first seared to brown its surface and enhance its flavor (through a process known as the Maillard reaction). If the food will not produce enough liquid of its own, a small amount of cooking liquid that often includes an acidic element, such as tomatoes, beer, or wine, is added to the pot, often with stock. Classically a braise is done with a relatively whole cut of meat, and the braising liquid will cover 2/3 of the product. The dish is cooked covered at a very low simmer until the meat is fork tender. Often the cooking liquid is finished to create a sauce or gravy.

Sometimes foods with high water content (particularly vegetables) can be cooked in their own juices and no extra liquid is required.

A successful braise intermingles the flavors of the foods being cooked and the cooking liquid. This cooking method dissolves collagen from the meat into gelatin, to enrich and add body to the liquid. Braising is economical, as it allows the use of tough and inexpensive cuts, and efficient, as it often employs a single pot to cook an entire meal.

It is possible to braise meat and vegetables in a pot on top of a grill. A gas or electric grill would be the best choice for what is known as barbecue-braising, or combining grilling directly on the surface and braising in a pot. To braise on a grill, put a pot on top of the grill, cover it, and let it simmer for a few hours. There are two advantages to barbecue-braising: the first is that this method now allows for browning the meat directly on the grill before the braising, and the second is that it also allows for glazing the meat with sauce and finishing it directly over the fire after the braising, effectively cooking the meat three times, which results in a soft-textured product that is very tender.

Familiar braised dishes include pot roast, beef stew, Swiss steak, chicken cacciatore, goulash, Carbonade Flamande, coq au vin, sauerbraten, beef bourguignon and Moroccan tajines, among others. Braising is also used extensively in the cuisines of Asia, particularly Chinese cuisine, where soy sauce is often the braising liquid.

Double Steaming

Double steaming, sometimes called double boiling, is a Chinese cooking technique to prepare delicate food such as bird's nest soup and shark fin soup. The food is covered with water and put in a covered ceramic jar and the jar is then steamed for several hours. This technique ensures there is no loss of liquid or moisture (its essences) from the food being cooked, hence it is often used with expensive ingredients like Chinese herbal medicines.

In Cantonese, double steaming is called dun. The meaning of the Chinese character for dun in Cantonese is different from that in Mandarin, because dun means to simmer or stew in Mandarin. This technique is also common in Fujian, a neighbouring province of Guangdong (Canton).

Cantonese cuisine is famous for its slow cooked soup. One famous dish of this kind is called the Winter melon urn. It is prepared by emptying the inside of a winter melon to make an urn. The outside of the winter melon is often carved with artistic patterns. The inside is then filled with soup ingredients such as Chinese cured ham, and several Chinese herbs. The whole urn completed with its original melon lid is double steamed for at least four hours. The flavour of the soup is soaked into the flesh of the melon. The whole melon and its content is brought to the dinner table. The soup is served by scooping out the liquid and the inside wall of the melon. In this case, the edible melon takes the place of the double steaming jar. This application is possible because winter melon has a waxy, and thus waterproof, rind. Winter melon is believed to be nourishing and it is seldom cooked with ingredients that are believed to be too yin or too yang.

There is another dessert dish called double steamed frog ovaries in a coconut, which is recommended for women. The Chinese medicinal ingredients (including hasma), spices, and rock sugar are placed inside a young coconut to soak in the original coconut juice. The filled coconut is then double steamed for several hours. The whole coconut is served whole at the table after dinner. The contents and the inside wall of the coconut are scooped out to be consumed.

Infusion

Infusion is the process of extracting chemical compounds or flavors from plant material in a solvent such as water, oil or alcohol, by allowing the material to remain suspended in the solvent over time (a process often called steeping). An infusion is also the name for the resultant liquid. The process of infusion is distinct from decoction, which involves boiling the plant material, or percolation, in which the water passes through the material (as in a coffeemaker).

An infusion is a very simple chemical process used with botanicals that are volatile and dissolve readily, or release their active ingredients easily, in water, oil or alcohol. The botanicals are typically dried herbs, flowers or berries. The liquid is typically boiled (or brought to another appropriate temperature) and then poured over the herb, which is then allowed to steep in the liquid for a period of time. The liquid may then be strained or the herbs otherwise removed from the liquid. Unless the infusion is to be consumed immediately, it may then be bottled and refrigerated for future use.

The amount of time the herbs are left in the liquid depends on the purpose for which the infusion is being prepared. Usually steeping for not more than 15 to 30 minutes, or until the mix cools, will create a beverage with optimal flavor. Steeping for a longer time typically results in a somewhat bitter-tasting infusion. Four (4) hours, however, is a more appropriate length of time for achieving herbal potency if health benefits are the priority. Quantities of the herb and liquid used will vary according to the herb or how strong the infusion is required to be. A common proportion used is 28 g (one ounce) of herb to 0.5 L (one pint) of liquid.

There have been several accessories and techniques for removing the steeped or left over products that were used to infuse liquids such as water, oil, or alcohol. The use of a metal steeper, which looks like a metal clamp. Tea infusers work as strainers and assist in removal of used herbs,leaves, etc., from over steeping or leaving residues. French presses are commonly used to infuse water with various teas and coffee. There are product lines such as Zing anything that are easy on the go beverage and oil infusers. Lastly, and most commonly used, the tea bag. Tea bags today are made with filter paper and filled with various tea flavors.

A common example of an infusion is tea, and many tisanes are prepared in the same way. Lemon, chamomile, senna, apple, ginger, rooibos, and a great many other plants are used individually or in combination. Herbal infusions in water and oil are both commonly used as herbal remedies. Coffee can also be made through infusion, but is more often made through percolation.

Plants with desirable flavors may be steeped in an edible oil or vinegar for an extended period; the infused oil or vinegar is often sold still containing the plant, and is then used as flavoring. Chiles, lemon, garlic, and many other plants may be used. There can be ambiguity in the labeling of these oils: for example, what is described as sesame oil may be oil extracted from sesame seeds, or another vegetable oil infused with sesame.

Poaching

Poaching is the process of gently simmering food in liquid, generally milk, stock or wine.

Poaching is particularly suitable for delicate food, such as eggs, poultry, fish and fruit, which might easily fall apart or dry out. For this reason, it is important to keep the heat low and to keep the poaching time to a bare minimum, which will also preserve the flavor of the food.

The poaching liquid is called court bouillon and a classical court bouillon consists of an acid (wine, lemon juice) and aromatics (bouquet garni and mirepoix). The liquid should ideally be around 160–85 °F, but when poaching chicken, it is vital that the chicken reach an internal temperature of at least 165 °F (74 °C) in the core, in order to be ingested safely.

Poached eggs are generally cooked in water and vinegar, fish in white wine, poultry in stock and fruit in red wine. Typically an egg is poached just to the point where the white is no longer runny and the yolk is beginning to harden around the edges.

Water is a relatively efficient conductor of heat, but it also has a fairly low limit to its maximum potential temperature (100 °C) at sea level. As such, it is a technique that applies itself to a broad spectrum of methods and results. It is used to regulate food at a low temperature for extended periods, as with Sous-vide. It is also used to rapidly raise the temperature of foods, as with Blanching. While it cannot achieve caramelization, which to many is very desirable, many find the delicate nuance of so-called "blanc" foods very pleasant.

Pressure Cooking

Pressure cooking is the process of cooking food, using water or other cooking liquid, in a sealed vessel — known as a pressure cooker, which does not permit air or liquids to escape below a pre-set pressure. Pressure cookers are used for cooking food quicker than conventional cooking methods, which also saves energy.

Pressure cookers heat food quickly because the internal steam pressure from the boiling liquid causes saturated steam (or "wet steam") to bombard and permeate the food. Thus, higher temperature water vapour (i.e., increased energy), which transfers heat more rapidly compared to dry air, cooks food very quickly.

Pressure cooking allows food to be cooked with greater humidity and higher temperatures than possible with conventional boiling or steaming methods. In an ordinary non-pressurised cooking vessel, the boiling point of water is 100 °C at standard pressure; the temperature of food is limited by the boiling point of water because excess heat causes boiling water to vaporize into steam. In a sealed pressure cooker, the boiling point of water increases as the pressure rises, resulting in superheated water. At a pressure of 15 psi (103 kPa) above atmospheric pressure, water in a pressure cooker can reach a temperature of up to 121 °C.

Pressure is created initially by boiling a liquid such as water or broth inside the closed pressure cooker. The trapped steam increases the internal pressure and temperature. After use, the pressure is slowly released so that the vessel can be safely opened.

Pressure cooking can be used to quickly simulate the effects of long braising or simmering. Almost any food which can be cooked in steam or water-based liquids can be cooked in a pressure cooker.

Variants

Several variants of the pressure cooker exist. An early pressure cooker, called a steam digester, was invented by Denis Papin, a French physicist, in 1679. Large pressure cookers are often called pressure canners in the United States because of their capacity to hold jars used in canning. Pressure canners are specifically designed for home canning, whereas ordinary pressure cookers are not recommended for canning due to the risk of botulism poisoning, because pressure canners hold heat and pressure for much longer than ordinary pressure cookers and these factors are a critical part of the total processing time required to destroy harmful microorganisms. An autoclave is a type of pressure cooker used by laboratories and hospitals to sterilize equipment. Pressure fryers are used for deep fat frying under pressure — note that ordinary pressure cookers are not suitable for pressure frying. In the food industry, pressure cookers are often referred to as retorts or canning retorts.

Design

Pressure cookers are typically made of aluminum (aluminium) or stainless steel. Aluminum pressure cookers may be stamped, polished, or anodized, but aluminum cookers are unsuitable for the dishwasher. Aluminium pressure cookers are cheaper, but the aluminium is reactive to acidic foods (changes the flavour) and less durable than stainless steel pressure cookers.

Higher-quality stainless steel pressure cookers are made with heavy, three-layer, or copper-clad bottoms (heat spreader) for uniform heating because stainless steel has lower thermal conductivity. Most modern stainless steel cookers are dishwasher safe, although some manufacturers may recommend washing by hand. Some pressure cookers have a non-stick interior.

In most models, a gasket or sealing ring, made from either rubber or silicone, forms a gas-tight seal that does not allow air or steam to escape between the lid and pan; more expensive models feature a metal-to-metal seal. Normally, the only way steam can escape is through a regulator on the lid while the cooker is pressurized. If the regulator becomes blocked, a safety valve provides a backup escape route for steam. A loose-fitting rubber plug in the lid, held in place by steam pressure, provides a simple safety valve. If the pressure exceeds design limits, the plug pops out to depressurize the pot.

To seal the gasket, some pressure cookers have a lid lock with flanges, similar to a bayonet-style lens mount, that works by placing the lid on the pot and twisting it about 30° to lock it in place. Contemporary designs of this style also have a pressure-activated interlock mechanism that prevents the lid from being removed while the cooker is pressurized.

Other cookers, particularly the larger types used for home canning, have over-sized oval lids. With such cookers, one inserts the lid at an angle, then turns the lid to align it with the pot opening on top because the lid is larger than the opening. A spring arrangement straddles the top of the cooker and holds the lid in place. While cooking, the internal pressurized steam keeps the lid tightly in place, preventing accidental removal.

Because of the forces that pressure cookers must withstand, they are usually heavier than conventional pots of similar size.

Generations

There are three generations of pressure cookers:

First generation

Also known as “old type” pressure cookers, these operate with a weight-modified or “jiggly” valve, which releases pressure during operation. Some people might consider them loud or very loud because the weight-modified valve operates similarly to the piston in a steam engine. They typically offer only one pressure level—with the exception of some newer “old style” pressure cookers that allow the operator to change the weight of the weight-modified valve.

Second generation

Also known as "new" or "latest generation" pressure cookers, these operate with a spring-loaded valve that is often hidden from view in a proprietary mechanism. This generation is characterized by two or more pressure settings. Some of these pressure cookers do not release any steam during operation (non-venting) and instead use a rising indicator with markings to show the pressure level. These only release steam when the pan is opened, or as a safety precaution if the heat source is not reduced enough when the pan reaches the required cooking pressure. Others use a dial that the operator can advance by a few clicks (which adjusts a spring underneath) to change the pressure setting or release pressure; these release steam during operation.

Electric pressure cookers

Called "third generation" pressure cookers by their manufacturers, these include an electric heat source that is automatically regulated to maintain the operating pressure. They also include a spring-loaded valve (as described above). Two or more pressure settings are available on this type of pressure cooker, along with features such as a timer and a setting to keep food warm. However, this pressure cooker type cannot be opened with a cold water quick-release method and should be operated with caution when releasing vapour through the valve, especially while cooking foamy foods and liquids (lentils, beans, grains, milk, gravy, etc.).

Pressure Settings

Most pressure cookers have a working pressure setting of 15 pounds per square inch (psi), sometimes expressed as "lb" or "lbs" (pounds), which equates to 103 kPa, 1.03 bar, or 1 kg per square centimetre (kg/cm^2 or kgf/cm^2) above atmospheric pressure. This standard was determined by the United States Department of Agriculture in 1917. At this pressure, water boils at 121 °C (250 °F) (described in vapour pressure of water article).

The higher temperature causes food to cook faster; cooking times can typically be reduced to 1/3 of the time for conventional cooking methods. The actual cooking time also depends on the pressure release method used after timing and the thickness and density of the food, since thicker foods take longer to cook. Meat joints and some other foods like sponge puddings and Christmas puddings are typically timed according to their weight. Frozen foods need extra cooking time to allow for thawing.

When pressure cooking at 15 psi, approximate cooking times are one minute for shredded cabbage and three minutes for fresh green beans, assuming that the pressure is immediately released after timing and not left to drop naturally.

If the pressure is allowed to slowly drop after timing, cooking times are even shorter. Examples of approximate cooking times using this method: potatoes cut into 1-inch pieces, seven minutes; a whole chicken, which easily fits into all but small pressure cookers, cooks in about twenty minutes. Stews can be cooked in under one hour, compared to several hours in a slow cooker or oven.

Some pressure cookers have a lower maximum pressure than the industry standard 15 psi or can be adjusted to different maximum pressures for some recipes; cooking times will increase or decrease accordingly. This is typically done by having different regulator weights or different pressure settings. If the pressure cooker is not capable of cooking at the industry standard 15 psi and the recipe is intended for cooking at 15 psi, the cooking time will need to be extended (e.g. the cooking time for 15 psi multiplied by 1.3), but this can overcook food. The manufacturer's instruction booklet should state the maximum operating pressure, possibly using different pressure units instead of the traditional "psi" rating (these pressure units are mentioned above). Below 15 psi, the alternative pressure units and values are: kPa below 100, bar below 1, $kgcm^2$ below 1. As an example, a pressure cooker with a maximum operating pressure of 0.8 bar equates to only 11.6 psi. Increasing the heat under pressure cookers designed for pressures below 15 psi will not make the food cook faster; it will only waste energy, possibly spoil food, increase the risk of boiling dry, and reduce the life of the gasket and other rubber parts. Virtually all electric pressure cookers operate below 15 psi.

Operation

Food is placed inside the pressure cooker with a small amount of water or other liquid (e.g., stock required for the recipe). Food is either cooked in the liquid or placed above the liquid to be steamed; the latter method prevents the transfer of flavours from the liquid. Steaming is done by using a suitable insert such as a metal steamer basket with a support trivet. The lid is closed, the pressure setting is selected, and the pressure cooker is then placed on a heat source such as a stove at the highest heat setting. For pressure cookers with a weight, the weight is placed over the steam vent pipe while steam is being emitted to ensure the air inside has escaped. Once the cooker reaches full pressure, the heat is lowered to maintain pressure; the timing specified in the recipe begins at this point. It takes several minutes for the pressure cooker to reach the selected pressure level. It can take as long as 10 minutes depending on: the quantity of food, the temperature of the food (cold or frozen food delays pressurisation), the amount of liquid, and the size of the pressure cooker.

Using boiling water from a kettle can quicken the heating process, unless the recipe calls for cold water. For example, boiled potatoes cook best when added to cold water because this prevents the potatoes from breaking up. Boiled potatoes also require the pressure to drop naturally after timing instead of being quickly released.

The newer generation pressure cookers, which have no weights, expel air from inside before reaching full pressure. A common mistake is for the user to start timing when the pop-up indicator rises, which happens when there is the slightest increase in pressure, instead of waiting for the cooker to reach the selected pressure level first. Some pressure cookers have markers on the pop-up indicator that show the pressure level. However, the typical pop-up indicator only shows that the cooker has pressure inside, which does not reliably signal that the cooker has reached the selected pressure. This indicator often acts as an interlock to prevent the lid from being opened while there is internal pressure. Manufacturers may use their own terminology for this indicator, such as calling it a locking indicator.

As the internal temperature rises, the pressure also rises until it reaches the design gauge pressure. In some models, a relief valve subsequently opens, releasing steam to prevent the pressure from rising any further. In other models, the pressure regulator weight begins levitating above its nozzle, allowing excess steam to escape. At this stage, the heat source should be reduced because heat is only needed to maintain pressure. If the heat source is too high, then energy is wasted, too much liquid may be released to maintain the nominal pressure, and reduces the life of the gasket/sealing ring by wear. On the other hand, if the heat source is too low, the food may be undercooked or pressure may be lost. As a rough guide, medium heat is the highest setting required on any domestic stove to maintain the pressure level throughout cooking time.

Recipes for foods using raising agents (e.g., steamed puddings) require gentle pre-steaming, without pressure, in order to activate the raising agents prior to cooking and achieve a light, fluffy texture. The water must bubble gently while pre-steaming to ensure that enough water will be available for the entire pressure cooking time, otherwise the pan will boil dry. If pre-steaming is omitted, it can result in a heavy, stodgy texture.

Since pressure cooking depends on the production of steam from water, the process cannot easily be used for cooking methods that produce little steam such as roasting, pan frying, or deep frying. However, Kentucky Fried Chicken restaurants use a combination of pressure cooking and frying, with special pressure fryers in which chicken juices supply the water. Cooking time is reduced substantially, but the breading texture is much softer (less crispy) than deep-fried chicken since moisture remains in the breading. Thick sauces do not contain enough liquid to vaporize and create pressure, so they usually burn onto the interior base of the pressure cooker after prolonged heating. Sauces should normally be thickened after pressure cooking.

Food Containers

Small containers such as plastic pudding containers, can be used in a pressure cooker, provided that the containers (and any covering used) can withstand temperatures of 130 °C and are not placed directly on the interior base. The containers can be used for cooking foods that are prone to burning on the base of the pressure cooker, which is directly in contact with the heat source. A lid for the container may be used, provided that the lid allows some steam to come into contact with the food and the lid is securely fitted (e.g. foil or greaseproof paper, pleated in the centre and tied securely with string). Cracked containers are not suitable. Cooking time is longer for covered containers because the food is not in direct contact with the steam. Since non-metal containers are poorer heat conductors, the type of container material stated in the recipe cannot be substituted without affecting the outcome. For example, if the recipe time is calculated using a stainless steel container and a plastic container is used instead, the food at the bottom of the container may be under-cooked. Liquid can be added inside the container when pressure cooking foods such as rice.

Pre-frying ingredients

The flavour of some foods, such as meat and onions, can be improved by gently cooking with a little pre-heated cooking oil, butter or other fat in the open pressure cooker over medium heat (unless the manufacturer advises against this) before pressure cooking. It is important not to overheat the empty pressure cooker and never heat the empty cooker with the lid and gasket in place. Overheating can cause warping and other damage. The pressure cooker should be allowed to cool briefly before adding liquid; otherwise some of the liquid will evaporate instantly, possibly leaving insufficient liquid for the entire pressure cooking time; if deglazing the pan, this has to be taken into account.

Pressure release methods

After cooking, there are three ways of releasing the pressure, either quickly or slowly, before the lid can be opened. Recipes for pressure cookers state which release method is required at the end of the cooking time for proper results. Failure to follow the recommendation may result in food that is under-cooked or over-cooked. Only one of these release methods is used after timing, as recommended in the recipe.To avoid opening the pressure cooker too often while cooking vegetables with varying cooking times, vegetables that take longer to cook can be cut into smaller pieces and vegetables that cook faster can be cut into thicker pieces.

Cold water quick release

This method is the fastest way of releasing pressure with portable pressure cookers; it is recommended to read the manufacturer's instruction book, as some may advise against the cold water release or require it to be performed differently. The cold water release method involves using slow running cold tap water, over the edge of the pressure cooker lid, being careful to avoid the steam vent or any other valves or outlets and never immersing the pressure cooker under water. It is most suitable for foods with short cooking times. It takes about 20 seconds for the cooker to cool down enough to lower the pressure so that it can be safely opened. This method is not suitable for electric pressure cookers.

Manual, normal, regular, or automatic release

This method is sometimes called a quick release, not to be confused with the cold water release. It involves the quick release of vapor by lifting (or removing) the valve, pushing a button, or turning a dial. It is most suitable to interrupt cooking to add food that cooks faster than what is already in the cooker. For example, since meat takes longer to cook than vegetables, it is necessary to add vegetables to stew later so that it will cook only for the last few minutes. Unlike the cold water release method, this release method does not cool down the pressure cooker. The user must release the steam with caution to avoid scalding injury. This release method is not suitable for foods that foam and froth while cooking; the hot contents might spray outwards due to the pressure released from the steam vent. This release method takes about two minutes to release the pressure before the lid can be opened.

Natural release

The natural release method allows the pressure to drop slowly; this is achieved by removing the pressure cooker from the heat source and allowing the pressure to lower without action. It takes approximately 10 to 15 minutes (possibly longer) for the pressure to disappear before the lid can be opened. On many pressure cookers, a coloured indicator pin will drop when the pressure has gone. This natural release method is recommended for foods that foam and froth during cooking, such as rice, pasta, or legumes, or for recipes such as steamed puddings with raising agents. The texture and tenderness of meat cooked in a pressure cooker may be improved by using the natural release method. The natural release method finishes cooking foods or recipes that have longer cooking times because the inside of the pressure cooker stays hot. The natural release method can be used with potatoes with skins to keep the skins intact if they have been deeply pierced before cooking. This method is not recommended for foods that require very short cooking times because the food will be overcooked. When using the natural release method for foods such as potatoes, the cooking time on the heat source is shorter than that of the other release methods. On the other hand, if the natural release method is not used, the cooking time will need to be extended by a few minutes to compensate.

Advantages

Foods cook much faster with pressure cooking than with other methods (except for small quantities in microwave ovens). Pressure cooking requires much less water than conventional boiling, so food can be ready sooner. Less energy is required than that of boiling, steaming, or oven cooking. Since less water or liquid has to be heated, the food reaches its cooking temperature faster. Using more liquid than necessary wastes energy because it takes longer to heat up; the liquid quantity is stated in the recipe. Pressure cookers can use much less liquid than the amount required for boiling or steaming in an ordinary saucepan. It is not necessary to immerse food in water. The minimum quantity of water or liquid used in the recipe to keep the pressure cooker filled with steam is sufficient. Because of this, vitamins and minerals are not leached (dissolved) away by water, as they would be if food were boiled in large amounts of water. Due to the shorter cooking time, vitamins are preserved relatively well during pressure cooking.

Several foods can be cooked together in the pressure cooker, either for the same amount of time or added later for different times. Manufacturers provide steamer baskets to allow more foods to be cooked together inside the pressure cooker.

Food is cooked at a temperature above the normal boiling point of water, killing most micro-organisms. A pressure cooker can be used as an effective sterilizer for jam pots, glass baby bottles, or for water while camping.

The pressure cooker speeds cooking considerably at high altitudes, where the lower atmospheric pressure reduces the boiling point of water. Lower water temperature reduces water's effectiveness for cooking or preparing hot drinks. The increased temperatures due to pressure cooking are also used to promote the Maillard reaction to develop more desirable flavor profiles that would not be obtainable using temperatures typical of boiling. The flavours are more concentrated in the higher temperature and sealed environment of the pressure cooker, so less seasoning is required.

Disadvantages

Pressure cookers are considerably more expensive than conventional saucepans of the same size. The additional gasket (sealing ring) requires special care when cleaning (e.g., not washed with kitchen knives), unlike a standard lid for a saucepan. Food debris must be cleaned from the gasket after every use to ensure a good seal. The gasket/sealing ring must be replaced if there are any signs of wear (e.g., a small split). If the gasket cannot provide an airtight seal and steam leaks around the lid, even when the gasket is fitted properly, the pressure cooker may not build up pressure inside. This can be caused by a defective gasket or if the rim of the pressure cooker is damaged in any way. The gasket should not be allowed to dry-out because this makes it harder to open and close the lid; however, the gasket can be smeared lightly with vegetable oil to help alleviate this problem (using too much lubrication can make the gasket swell and it may not fit properly).

In order to inspect the food, the pressure cooker needs to be opened, which halts the cooking process. With a conventional saucepan, this can be done in a matter of seconds by visually inspecting the food. As a result, accurate timing is essential to the recipe and the recommended cooking time for the food must be monitored e.g. with an audible timer.

The increased weight of conventional pressure cookers makes them unsuitable for applications in which saving weight is a priority, such as camping. However, small, lightweight pressure cookers are available for mountain climbers.

A minimum quantity of liquid is required to create and maintain pressure, as indicated in the manufacturer's instruction manual. More liquid is required for longer cooking times. This is not desirable for food requiring much less liquid, but recipes and books for pressure cookers take this into account.

Safety Features

Early pressure cookers equipped with only a primary safety valve risked explosion from food blocking the release valve. On modern pressure cookers, food residues blocking the steam vent or the liquid boiling dry would trigger additional safety devices. Modern pressure cookers sold from reputable manufacturers have sufficient safety features to prevent the pressure cooker itself from exploding. When excess pressure is released by a safety mechanism, debris of the food being cooked may also be ejected with the steam — sometimes forcefully with some noise. This can be avoided if the pressure cooker is regularly cleaned and maintained in accordance with the manufacturer's instructions and never overfilled with food and/or liquid. A pressure cooker should never be filled to more than 2/3 its interior height with solid food, 1/2 full for liquids and foods that foam and froth (e.g., rice, pasta), and no more than 1/3 full for pulses (e.g. lentils).

Modern pressure cookers typically have two or three redundant safety valves and some additional safety features, such as an interlock lid that prevents the user from opening the lid when the internal pressure exceeds atmospheric pressure, preventing accidents from a sudden release of hot liquid, steam and food. If any safety mechanisms are not correctly in place, the cooker will not pressurize the contents. Pressure cookers should be operated only after reading the instruction manual because each model has slightly different safety mechanisms.

For first generation pressure cookers with a weighted valve or "jiggler", the primary safety valve or regulator is usually a weighted stopper, commonly called "the rocker" or "vent weight". This weighted stopper is lifted by the steam pressure, allowing excess pressure to be released. There is a backup pressure release mechanism that releases pressure quickly if the primary pressure release mechanism fails (e.g., food jams the steam discharge path). One such method is a hole in the lid that is blocked by a low melting point alloy plug and another is a rubber grommet with a metal insert at the center. At a sufficiently high pressure, the grommet will distort and the insert will blow out of its mounting hole to release pressure. If the pressure continues to increase, the grommet itself will blow out to release pressure. These safety devices usually require replacement if they are activated by excess pressure. Newer pressure cookers may have a self-resettable spring device, fixed onto the lid, that releases excess pressure.

For second generation pressure cookers with selectable pressure and (often hidden) spring valve operation, a common safety feature is the gasket, which expands and releases excess pressure downward between the lid and the pot. This release of excess pressure is forceful and sufficient to extinguish the flame of a gas stove.

Science of Pressure Cooking

The boiling point of water (and water-based liquids) is determined by the ambient pressure. Pressure cookers always require liquid in order to cook food under pressure. At atmospheric pressure and at sea level, the boiling point of water is 100°C and excess heat only increases the rate at which water evaporates into steam vapour; more heat does not increase the temperature of the water. At higher altitudes above sea level, the atmospheric pressure is lower and thus the boiling point of water is lower, because the lower pressure pushing on the water makes it easier for the water molecules to escape to the surface compared to higher pressure. Inside a pressure cooker, once the water (liquid) is boiling and the steam is trapped, the pressure from the trapped steam increases and this pushes on the liquid, which increases its boiling point, because it becomes harder for the water molecules to escape from the surface as the pressure increases on it. The heat applied to the liquid by the heat source continues to create more steam pressure and the extra heat also raises the temperature of the liquid under this increased pressure. Both the liquid and steam are at the same temperature. Once the selected pressure level is reached, the pressure regulator on the lid indicates this and now the heat source can be lowered to maintain that pressure level and save energy, since extra heat will not increase the temperature of the liquid if the pressure is not allowed to rise — excess pressure will only escape as fast-flowing steam from the lid.

Steam and liquids transfer heat more rapidly than dry air. As an example, the hot air inside an oven at, say 200°C , won't immediately burn your skin, but the wet steam from a boiling kettle at 100°C will scald your skin almost instantly and 'feel' hotter, despite the steam in the kettle being at a lower temperature than the air inside a hot oven. Since steam and liquids heat substances faster, the pressure cooker can cook food quicker — under pressure — compared to ordinary cooking methods.

Use in food detoxification

Some food toxins can be reduced by pressure cooking. A Korean study of aflatoxins in rice (associated with Aspergillus fungus) showed that pressure cooking was capable of reducing aflatoxin concentrations to 12-22% of the amount in the uncooked rice. Pressure cookers are not guaranteed to destroy all harmful microorganisms in food, especially when used for short periods of time.

Foods unsuitable for pressure cooking

Some foods are not recommended for pressure cooking in liquid. Such foods could expand too much, froth, and sputter, which can result in blocking the steam vent. Examples include macaroni, cranberries, and cereals such as oatmeal. Dumplings should not be cooked under pressure, otherwise the dumplings will expand, block the steam vent, and trigger the safety devices that prevent over-pressure. Dumplings can be added to a stew already cooked in the pressure cooker and gently simmered without the lid on.

SIMMERING

Simmering is a food preparation technique in which foods are cooked in hot liquids kept at or just below the boiling point of water (which is 100 °C or 212 °F at average sea level air pressure), but higher than poaching temperature. To keep a pot simmering, one brings it to a boil and then reduces the heat to a point where the formation of bubbles has all but ceased, typically a water temperature of about 94°C.

Simmering ensures gentler treatment than boiling to prevent food from toughening and/or breaking up. Simmering is usually a rapid and efficient method of cooking. Food that has simmered in milk or cream instead of water is referred to as creamed. The appropriate simmering temperature is a topic of debate among chefs, with some contending that a simmer is as low as 82°C.

Smothering

Smothering meat, seafood or vegetables is a cooking technique used in both Cajun and Creole cuisines of Louisiana. The technique involves cooking in a covered pan over low heat with a small amount of liquid. The meat dishes cooked in this fashion are typically served over boiled or steamed white rice as a rice and gravy, while the vegetables are typically served as side dishes.

his method of food preparation involves using the Maillard reaction to "brown" the meat or vegetables and then deglazing with stock or water and simmering the mixture over low heat for an extended period of time. It is often done in a cast iron pot or dutch oven, so the heat can be evenly applied and distributed.

Meat

A large variety of meats are "smothered" in South Louisiana cuisine, including both domestic animals and wild game. Domestic animals cooked in this fashion include chicken, domestic duck, pork, beef (including such organs as the liver), and domestic rabbit. Wild game commonly cooked in this fashion include squirrel, rabbit, nutria rat, feral pig, woodcock, wild duck, and venison. Originally a dish made from cheap cuts of meat favored by farmers and laborers, popular versions of the dish such as "smothered steak" and "smothered pork roast" are served throughout Acadiana at local "plate lunch houses". Raised on Rice and Gravy, a 2009 documentary film by Conni Castille and Allison Bohl, chronicles the prevalence of the dish at local plate lunch houses and its enduring popularity in local cuisine.

Shellfish

In French, the word "étouffée" means "smothered". Étouffée can be made using different shellfish, the most popular version of the dish being Crawfish

Étouffée, although shrimp is also used. Originally étouffée was a popular dish in the Acadiana area surrounding Breaux Bridge. In the late twentieth century a waiter at the popular Bourbon Street restaurant Galatoire's brought the dish in to his employer to try. The dish was added to their menu and other restaurants in the city soon followed, with the dish gaining in popularity with locals and tourists alike. Many Cajun restaurant owners claim that étouffée is the most popular dish on their menus.

Vegetables

Varieties of vegetables cooked by smothering include cabbage, okra, potatoes and corn. The vegetables are kept from burning by the addition of animal fats or oils, or the addition of meat products such as salt pork or andouille.

Steaming

Steaming is a method of cooking using steam. Steaming is considered a healthy cooking technique and capable of cooking almost all kinds of food.

Steaming works by boiling water continuously, causing it to vaporize into steam; the steam then carries heat to the nearby food, thus cooking the food. The food is kept separate from the boiling water but has direct contact with the steam, resulting in a moist texture to the food. This differs from double boiling, in which contact with steam is undesired.

Such cooking is most often done by placing the food into a steamer, which is typically a circular container made of metal or bamboo. The steamer usually has a lid that is placed on the top of the container during cooking to allow the steam to cook the food. When a steamer is unavailable, a wok filled less than half with water is a replacement by placing a metal frame made of stainless steel in the middle of the wok. Some modern home microwave ovens include the structure to cook food by steam vapor produced in a separate water container, providing a similar result to being cooked by fire.

Benefits

Overcooking or burning food is easily avoided when steaming it. Individuals looking to reduce their fat intake may prefer steaming to other methods which require cooking oil. Steaming also results in a more nutritious food than boiling because fewer nutrients are leached away into the water, which is usually discarded. A 2007 USDA comparison between steaming and boiling vegetables shows the most affected nutrients are folic acid and vitamin C. When compared to raw consumption, steaming reduces folic acid by 15%, and boiling reduces it by 35%. Steaming reduces vitamin C by 15%, and boiling reduces it by 25%. Phenolic compounds with antioxidant properties have been found to retain significantly better through steaming than through boiling or microwaving. Most other nutrients are reduced by a similar amount by both methods of cooking.

Food by Steaming

In China, the Chinese have used steamers for at least 3000 years or more, with early steamers being made in stoneware and originating in the province of Yunnan. From the eighth century, thin cypress strips were used to make steamers and today they are constructed from bamboo with slatted bases.

In Western cooking, steaming is most often used to cook vegetables - it is rarely used to cook meats. In Chinese cuisine, vegetables are mostly stir fried or blanched and seldom steamed. Seafood and meat dishes are steamed. For example: steamed whole fish, steamed crab, steamed pork spare ribs, steamed ground pork or beef, steamed chicken, steamed goose, etc. Other than meat dishes, rice can be steamed too, although in Chinese this is rarely referred to as "steaming" but rather simply as "cooking." Wheat foods are steamed as well. Examples include buns, Chinese steamed cakes etc. Steamed meat dishes (except fish and some dim sum) are less common in Chinese restaurants than in traditional home cooking because meats usually require longer cooking times to steam than to stir fry. Commercially sold frozen foods (such as dim sum) used to have instructions to reheat by steaming, until the rise in popularity of home microwave ovens which have considerably shorter cooking times.

Steeping

Steeping is the soaking in liquid (usually water) of a solid so as to extract flavours or to soften it. Some teas are prepared for drinking by steeping the leaves in heated water to release the flavour and nutrients. Herbal teas may be prepared by decoction, infusion, or maceration. Some solids are soaked to remove an ingredient, such as salt from smoked ham or salted cod, where the solvent is not the desired product.

One example is the steeping of corn (or maize), part of the milling process. As described by the US Corn Refiners Association, harvested kernels of corn are cleaned and then steeped in water at a temperature of 50 °C for 30 to 40 hours. In the process their moisture content rises from 15% to 45% and their volume more than doubles. The gluten bonds in the corn are weakened and starch is released. The corn is then ground to break free the germ and other components, and the water used (steepwater), which has absorbed various nutrients, is recycled for use in animal feeds.

Stewing

A stew is a combination of solid food ingredients that have been cooked in liquid and served in the resultant gravy. Ingredients in a stew can include any combination of vegetables (such as carrots, potatoes, beans, peppers and tomatoes, etc.), meat, especially tougher meats suitable for slow-cooking, such as beef. Poultry, sausages, and seafood are also used. While water can be used as the stew-cooking liquid, wine, stock, and beer are also common. Seasoning and flavourings may also be added. Stews are typically cooked at a relatively low temperature (simmered, not boiled), allowing flavors to mingle.

Stewing is suitable for the least tender cuts of meat that become tender and juicy with the slow moist heat method. This makes it popular in low-cost cooking. Cuts having a certain amount of marbling and gelatinous connective tissue give moist, juicy stews, while lean meat may easily become dry.

Stews may be thickened by reduction or with flour, either by coating pieces of meat with flour before searing, or by using a roux or beurre manié, a dough consisting of equal parts of butter and flour. Thickeners like cornstarch or arrowroot may also be used.

Stews are similar to soups, and in some cases there may not be a clear distinction between the two. Generally, stews have less liquid than soups, are much thicker and require longer cooking over low heat. While soups are almost always served in a bowl, stews may be thick enough to be served on a plate with the gravy as a sauce over the solid ingredients.

Stews have been made since ancient times. Herodotus says that the Scythians (8th to 4th centuries BC) "put the flesh into an animal's paunch, mix water with it, and boil it like that over the bone fire. The bones burn very well, and the paunch easily contains all the meat once it has been stripped off. In this way an ox, or any other sacrificial beast, is ingeniously made to boil itself."

Amazonian tribes used the shells of turtles as vessels, boiling the entrails of the turtle and various other ingredients in them. Other cultures used the shells of large mollusks (clams etc.) to boil foods in. There is archaeological evidence of these practices going back 8,000 years or more.

There are recipes for lamb stews and fish stews in the Roman cookery book Apicius, believed to date from the 4th century AD. Le Viandier, one of the oldest cookbooks in French, written by the French chef known as Taillevent, has ragouts or stews of various types in it.

Hungarian Goulash dates back to the 9th century Magyar shepherds of the area, before the existence of Hungary. Paprika was added in the 18th century.

Types of Stew

In meat-based stews, white stews, also known as blanquettes or fricassées, are made with lamb or veal that is blanched, or lightly seared without browning, and cooked in stock. Brown stews are made with pieces of red meat that are first seared or browned, before a browned mirepoix, sometimes browned flour, stock and wine are added. These choices of stew are all unique to the individuals' personal stew preference.

Vacuum Flask Cooking

Vacuum flask cooking was introduced to the Asian market in the mid-1990s. The vacuum cooker, often called a thermal cooker in English, is a stainless steel vacuum flask. The flasks come in various sizes ranging from 20 to 40 centimetres (8 to 16 in) in diameter and 25 centimetres (10 in) tall. A removable pot, with handle and lid, fits inside the vacuum flask. The pot and contents are heated to cooking temperature, and then sealed in the flask. The flask simply reduces heat loss to a minimum, so that the food remains at cooking temperature for a long time, and cooks without continued heating. Note that the food is not cooked in a vacuum. It is cooked inside a vacuum flask. The hollow evacuated wall of the cooker thermally insulates its contents from the environment, so they remain hot for several hours.

Vacuum flasks appeal to Cantonese cooks because many Cantonese dishes require prolonged braising or simmering. When these cookers were first introduced in the US, they sold very quickly in the larger Asian supermarkets. The slow cooker is used for a similar purpose; but instead of minimising heat loss, sufficient heat is applied to the non-insulated slow cooker to maintain a steady temperature somewhat below the boiling point of water. A slow cooker allows any desired cooking time; the more energy-efficient vacuum flask must cook within the time taken for the food to cool below cooking temperature.

The historical equivalent of the vacuum cooker is the haybox, nowadays perhaps using more modern insulating material than the original hay or straw. This works on the same principle but has much poorer heat retention.

The pot is filled with food and heated to cooking temperature, usually to boiling. It is then sealed inside the vacuum flask for several hours. The flask minimises heat loss, keeping the food hot enough to continue cooking.

Some flasks also come with 2 or 3 smaller pots so you can cook 2+ dishes at once.

If a large part of the cooking time is spent at temperatures lower than 60 °C (as when the contents of the cooker are slowly cooling over a long period), a danger of food poisoning due to bacterial infection, or toxins produced by multiplying bacteria, arises. It is essential to heat food sufficiently at the outset of vacuum cooking. 60 °C throughout the dish for 10 minutes is sufficient to kill most pathogens of interest, effectively pasteurizing the dish. The best practice is to bring briefly to a rolling boil then put the pot in the flask. This keeps it hottest longest.

Frying

Frying is the cooking of food in oil or another fat, a technique that originated in ancient Egypt around 2500 BC. Chemically, oils and fats are the same, differing only in melting point, but the distinction is only made when needed. Foods can be fried in a variety of fats, including lard, vegetable oil, rapeseed oil and olive oil. To fry in olive or vegetable oil is sometimes seen as healthier than doing so in lard, because the chief fat in olive oil is Monounsaturated fat, not saturated fat. In commerce, many fats are called oils by custom, e.g. palm oil and coconut oil, which are solid at room temperature. A variety of foods may be fried, including the Potato chip, bread, eggs and foods made from eggs, such as omelettes or pancakes.

Fats can reach much higher temperatures than water at normal atmospheric pressure. Through frying, one can sear or even carbonize the surface of foods while caramelizing sugars. The food is cooked much more quickly and has a characteristic crispness and texture. Depending on the food, the fat will penetrate it to varying degrees, contributing richness, lubricity, and its own flavor, as well as calories.

Frying techniques vary in the amount of fat required, the cooking time, the type of cooking vessel required, and the manipulation of the food. Sautéing, stir frying, pan frying, shallow frying, and deep frying are all standard frying techniques.

Sautéing and stir-frying involve cooking foods in a thin layer of fat on a hot surface, such as a frying pan, griddle, wok, or sauteuse. Stir frying involves frying quickly at very high temperatures, requiring that the food be stirred continuously to prevent it from adhering to the cooking surface and burning.

Shallow frying is a type of pan frying using only enough fat to immerse approximately one-third to one-half of each piece of food; fat used in this technique is typically only used once. Deep-frying, on the other hand, involves totally immersing the food in hot oil, which is normally topped up and used several times before being disposed. Deep-frying is typically a much more involved process, and may require specialized oils for optimal results.

Deep frying is now the basis of a very large and expanding worldwide industry. Fried products have consumer appeal in all age groups and in virtually all cultures, and the process is quick, can easily be made continuous for mass production, and the food emerges sterile and dry, with a relatively long shelf life. The end products can then be easily packaged for storage and distribution. Examples are potato chips, french fries, nuts, doughnuts, instant noodles, etc.

Deep Frying

Deep frying is a cooking method in which food is submerged in hot fat e.g. oil. This is normally performed with a deep fryer or chip pan; industrially, a pressure fryer or vacuum fryer may be used.

Deep frying is classified as a dry cooking method because no water is used. Due to the high temperature involved and the high heat conduction of oil, it cooks food extremely quickly.

If performed properly, deep-frying does not make food excessively greasy, because the moisture in the food repels the oil. The hot oil heats the water within the food, steaming it; oil cannot go against the direction of this powerful flow because (due to its high temperature) the water vapor pushes the bubbles toward the surface. As long as the oil is hot enough and the food is not immersed in the oil for too long, oil penetration will be confined to the outer surface. However, if the food is cooked in the oil for too long, much of the water will be lost and the oil will begin to penetrate the food. The correct frying temperature depends on the thickness and type of food, but in most cases it lies between 175-190 °C.

Oil deterioration

Overheating or over-using the frying oil leads to formation of rancid-tasting products of oxidation, polymerization, and other deleterious, unintended or even toxic compounds such as acrylamide (from starchy foods). Deep-frying under vacuum helps to significantly reduce acrylamide formation, but this process is not widely used in the food industry due to the high investment cost involved.

Some useful tests and indicators of excessive oil deterioration are the following:

- Sensory: Darkening, smoke, foaming, thickening, rancid taste and unpleasant smell when heating.
- Laboratory: Acidity, anisidine value, viscosity, total polar compounds, polymeric triglycerides.
- Instruments that indicate total polar compounds, currently the best single gauge of how deep-fried an object is, are available with sufficient accuracy for restaurant and industry use.

Disadvantages

Deep frying produces large amounts of waste oil, which must be properly disposed. Waste oil is increasingly being recycled into biodiesel.

Cooking oil is flammable, and fires may be caused by it igniting at too high a temperature. Further, attempts to extinguish an oil fire with water cause an extremely dangerous condition, a boilover, as it causes the water to flash into steam due to the high heat of the oil, in turn sending the burning oil in all directions and thus aggravating the fire. This is the leading cause of house fires in the United Kingdom. Instead, oil fires must be extinguished via fire extinguisher or by smothering. Other means of extinguishing an oil fire include application of dry powder (e.g., baking soda, salt) or fire fighting foam. Most commercial deep fryers are equipped with automatic fire suppression systems using foam.

Spilled hot cooking oil can also cause severe, even fatal, burns. The higher temperatures and tendency of oil to stick to the skin make spilled hot cooking oil far more dangerous than spilled hot water. Children are particularly at risk, as they can accidentally pull a pot off the stove onto themselves.

Cooking with oil also tends to make a mess, with oil splatters on all adjacent surfaces, and oil vapors condensing on more distant surfaces. More detergent is also needed to clean cooking pots and utensils coated in oil.

Hot Salt Frying

Hot salt frying is a cooking technique used by street-side food vendors in China/India. Coarse sea salt is placed in a large wok and heated to a high temperature. Dry food items, such as eggs in shell, are buried in the hot salt and occasionally turned with a spatula. This technique is also seen in India, where street vendors sell shelled peanuts or popcorn cooked in salt heated in an iron wok.

Pan Frying

Pan frying is a form of frying characterized by the use of minimal cooking oil or fat (compared to shallow frying or deep frying); typically using just enough oil to lubricate the pan (although, in the case of a greasy food such as bacon, no oil or fats may be needed). As a form of frying, pan frying relies on oil as the heat transfer medium and on correct temperature and time to retain the moisture in the food. Because of the partial coverage, the food must be flipped at least once to cook both sides.

Generally, a shallower cooking vessel is used for pan frying than deep frying. Using a deep pan with a small amount of oil, butter or bacon grease does reduce spatter. A denser cooking vessel is better than a less dense pan because that mass will improve temperature regulation. An electric skillet can be used analogously to an electric deep fryer and many of these devices have a thermostat to keep the liquid at the desired temperature.

The fastest and best way to pan fry is to flip the cooked item very often, about every 15–30 seconds. This will over cook the outside less and get the center cooked faster.

Pressure Frying

Pressure frying is a variation on pressure cooking where meat and cooking oil are brought to high temperatures while pressure is held high enough to cook the food more quickly. This leaves the meat very hot and juicy. A receptacle used in pressure frying is known as a pressure fryer.

The process is most notable for its use in the preparation of fried chicken in many commercial fried chicken restaurants.

Pressure frying is mostly done in industrial kitchens. Ordinary pressure cookers are generally unsuitable for pressure frying, because they are typically designed for a maximum temperature around 121°C whereas oil can reach temperatures well in excess of 160 °C which may damage the gasket in an ordinary pressure cooker causing it to fail. Attempting to pressure fry using a first generation pressure cooker can be very dangerous. Second generation cookers are somewhat safer, but can still be dangerous if proper precautions, like opening fill limits, are not taken.

Pressure fryers operate at a lower pressure than pressure cookers. The gasket in a pressure cooker's lid can be melted by boiling oil, which can result in burn injuries to anyone in the vicinity.

Sautéing

Sautéing is a method of cooking food, that uses a small amount of oil or fat in a shallow pan over relatively high heat. Ingredients are usually cut into pieces or thinly sliced to facilitate fast cooking. The primary mode of heat transfer during sautéing is conduction between the pan and the food being cooked. Food that is sautéed is browned while preserving its texture, moisture and flavor. If meat, chicken, or fish is sautéed, the sauté is often finished by deglazing the pan's residue to make a sauce.

Sautéing is often confused with pan frying, in which larger pieces of food (for example, chops or steaks) are cooked quickly, and flipped onto both sides. Some cooks make a distinction between the two based on the depth of the oil used, while others use the terms interchangeably. Sautéing differs from searing in that searing only browns the surface of the food. Olive oil or clarified butter are commonly used for sautéing, but most fats will do. Regular butter will produce more flavor but will burn at a lower temperature and more quickly than other fats due to the presence of milk solids, so clarified butter is more fit for this use.

Performing a Sauté

In a sauté, all the ingredients are heated at once, and cooked quickly. To facilitate this, the ingredients are rapidly moved around in the pan, either by the use of a utensil, or by repeatedly jerking the pan itself. A sauté pan must be large enough to hold all of the food in one layer, so steam can escape - which keeps the ingredients from stewing, and promotes the development of fond. Most pans sold specifically as sauté pans have a wide flat base and low sides, to maximize the surface area available for heating. The low sides allow quick evaporation and escape of steam. While skillets typically have flared or rounded sides, sauté pans typically have straight, vertical sides - this keeps the ingredients from escaping as the pan is jerked or stirred.

Only enough fat to lightly coat the bottom of the pan is needed for sautéing; too much fat will cause the food to fry rather than just to slide, and may interfere with the development of fond. The food is spread across the hot fat in the pan, and left to brown, turning or tossing frequently for even cooking. The sauté technique involves gripping the handle of the sauté pan firmly, and using a sharp elbow motion to rapidly jerk the pan back toward the cook, repeating as necessary to ensure the ingredients have been thoroughly jumped. Tossing or stirring the items in the pan by shaking the pan too often, however, can cause the pan to cool faster and make the sauté take longer.

Stir Frying

Stir frying is a pair of Chinese cooking techniques for preparing food in a wok: chao and bào. The term stir-fry was introduced into the English language by Buwei Yang Chao, in her book How to Cook and Eat in Chinese, to describe the chao technique. The two techniques differ in their speed of execution, the amount of heat used, and the amount of tossing done to cook the food in the wok. Cantonese restaurant patrons judge a chef's ability to perform stir frying by the "wok hei" produced in the food. This in turn is believed to display their ability to bring out the qi of the wok and essence of the food cooking.

Chao technique

The chao technique is similar to the Western technique of sautéing. A traditional round-bottom cast iron or carbon steel pan called a wok is heated to a high temperature. A small amount of cooking oil is then poured down the side of the wok (a traditional expression in China regarding this is "hot wok, cold oil"), followed by dry seasonings (including ginger and garlic), then at the first moment the seasonings can be smelled, meats are added and agitated. Once the meat is seared, vegetables along with liquid ingredients (for example often including premixed combinations of soy sauce, vinegar, wine, salt, sugar, and cornstarch) are added. The wok then may be covered for a moment so the water in the liquid ingredients can warm up the new ingredients as it steams off. To keep the meat juicy, usually a cook would take the seared meat out before vegetables are added, and put the meat back right before vegetables are done. In some dishes, or if the cooking conditions are inadequate, different components may be stir fried separately before being combined in the final dish (if, for example, the chef desires the taste of the stir fried vegetables and meats to remain distinct).

The food is stirred and tossed out very quickly using wooden or metal cooking utensils. Some chefs will lift the wok to the side to let the flame light the oil or add a dash of wine spirit to give the food extra flavor. Using this method, many dishes can be cooked extremely quickly (within a minute).

Some dishes that require more time are cooked by adding a few dashes of water after the stirring. Then the wok is covered with a lid. As soon as steam starts to come out from under the lid, the dish is ready. In this case, the food is stir fried on high heat for flavor and then steamed to ensure that it is fully cooked.

Bao technique

The wok is heated to a dull red glow. With the wok hot, the oil, seasonings, and meats are added in rapid succession with no pause in between. The food is continually tossed, stopping for several seconds only to add other ingredients such as various seasonings, broths, or vegetables. When the food is deemed to be cooked it is poured and ladled out of the wok. The wok must then be quickly rinsed to prevent food residues from charring and burning to the wok bottom because of residual heat.

The main ingredients are usually cut to smaller pieces to aid in cooking. As well, a larger amount of cooking fat with a high smoke point, such as lard and/or peanut oil, is often used in bao.

Smoking

Smoking is the process of flavoring, cooking, or preserving food by exposing it to the smoke from burning or smoldering plant materials, most often wood. Meats and fish are the most common smoked foods, though cheeses, vegetables, and ingredients used to make beverages such as whisky, smoked beer, and lapsang souchong tea are also smoked.

In Europe, alder is the traditional smoking wood, but oak is more often used now, and beech to a lesser extent. In North America, hickory, mesquite, oak, pecan, alder, maple, and fruit-tree woods, such as apple, cherry, and plum, are commonly used for smoking. Other fuels besides wood can also be employed, sometimes with the addition of flavoring ingredients. Chinese tea-smoking uses a mixture of uncooked rice, sugar, and tea, heated at the base of a wok. Some North American ham and bacon makers smoke their products over burning corncobs. Peat is burned to dry and smoke the barley malt used to make whisky and some beers. In New Zealand, sawdust from the native manuka (tea tree) is commonly used for hot smoking fish. In Iceland, dried sheep dung is used to cold-smoke fish, lamb, mutton, and whale.

Historically, farms in the Western world included a small building termed the smokehouse, where meats could be smoked and stored. This was generally well-separated from other buildings both because of the fire danger and because of the smoke emanations.

It is believed that the smoking of food dates back to the time of primitive cavemen. As caves or simple huts lacked chimneys, these dwellings could become very smoky. The primitive men would often hang meat up to dry, and it is presumed that at some point they became aware that meat that was stored in smoky areas acquired a different flavor and was better preserved than meat that simply dried out. Over time this process was combined with pre-curing the food in salt or salty brines, resulting in a remarkably effective preservation process that was adapted or developed independently by numerous cultures around the world. Until the modern era, smoking was of a more "heavy duty" nature as the main goal was to preserve the food. Large quantities of salt were used in the curing process and smoking times were quite long, sometimes involving days of exposure.

The advent of modern transportation made it easier to transport food products over long distances and the need for the time and material intensive heavy salting and smoking declined. Smoking came to be seen more as way to flavor than to preserve food. In 1939 a device called the Torry Kiln was invented at the Torry Research Station in Scotland. The kiln allowed for uniform mass-smoking and is considered the prototype for all modern large-scale commercial smokers. Although refinements in technique and advancements in technology have made smoking much easier, the basic steps involved remain essentially the same today as they were hundreds if not thousands of years ago.

Types

Cold smoking can be used as a flavor enhancer for items such as chicken breasts, beef, pork chops, salmon, scallops, and steak. The item can be cold smoked for just long enough to give some flavor. Some cold smoked foods are baked, grilled, roasted, or sautéed before eating. Smokehouse temperatures for cold smoking are typically done between 20 to 30 °C (68 to 86 °F). In this temperature range, foods take on a smoked flavor, but remain relatively moist. Cold smoking does not cook foods. Meats should be fully cured before cold smoking.

Hot smoking exposes the foods to smoke and heat in a controlled environment. Although foods that have been hot smoked are often reheated or cooked, they are typically safe to eat without further cooking. Hams and ham hocks are fully cooked once they are properly smoked. Hot smoking occurs within the range of 52 to 80 °C (126 to 176 °F). Within this temperature range, foods are fully cooked, moist, and flavorful. If the smoker is allowed to get hotter than 185 °F (85 °C), the foods will shrink excessively, buckle, or even split. Smoking at high temperatures also reduces yield, as both moisture and fat are "cooked" away.

Smoke roasting or smoke baking refers to any process that has the attributes of smoking combined with either roasting or baking. This smoking method is sometimes referred to as "barbecuing", "pit baking", or "pit roasting". It may be done in a smoke roaster, closed wood-fired masonry oven or barbecue pit, any smoker that can reach above 250 °F (121 °C), or in a conventional oven by placing a pan filled with hardwood chips on the floor of the oven so the chips smolder and produce a smokebath. However, this should only be done in a well-ventilated area to prevent carbon monoxide poisoning.

Wood Smoke

Hardwoods are made up mostly of three materials: cellulose, hemicellulose, and lignin. Cellulose and hemicellulose are the basic structural material of the wood cells; lignin acts as a kind of cell-bonding glue. Some softwoods, especially pines and firs, hold significant quantities of resin, which produces a harsh-tasting soot when burned; these woods are not often used for smoking.

Cellulose and hemicellulose are aggregate sugar molecules; when burnt, they effectively caramelize, producing carbonyls, which provide most of the color components and sweet, flowery, and fruity aromas. Lignin, a highly complex arrangement of interlocked phenolic molecules, also produces a number of distinctive aromatic elements when burnt, including smoky, spicy, and pungent compounds such as guaiacol, phenol, and syringol, and sweeter scents such as the vanilla-scented vanillin and clove-like isoeugenol. Guaiacol is the phenolic compound most responsible for the "smokey" taste, while syringol is the primary contributor to smokey aroma. Wood also contains small quantities of proteins, which contribute roasted flavors. Many of the odor compounds in wood smoke, especially the phenolic compounds, are unstable, dissipating after a few weeks or months.

A number of wood smoke compounds act as preservatives. Phenol and other phenolic compounds in wood smoke are both antioxidants, which slow rancidification of animal fats, and antimicrobials, which slow bacterial growth. Other antimicrobials in wood smoke include formaldehyde, acetic acid, and other organic acids, which give wood smoke a low pH—about 2.5. Some of these compounds are toxic to people as well, and may have health effects in the quantities found in cooking applications.

Since different species of trees have different ratios of components, various types of wood do impart a different flavor to food. Another important factor is the temperature at which the wood burns. High-temperature fires see the flavor molecules broken down further into unpleasant or flavorless compounds. The optimal conditions for smoke flavor are low, smoldering temperatures between 570 and 750 °F (299 and 399 °C). This is the temperature of the burning wood itself, not of the smoking environment, which uses much lower temperatures. Woods that are high in lignin content tend to burn hot; to keep them smoldering requires restricted oxygen supplies or a high moisture content. When smoking using wood chips or chunks, the combustion temperature is often raised by soaking the pieces in water before placing them on a fire.

Types of Smokers

There are a few basic types of smoker designs, each with their own advantages and disadvantages.

Offset smokers

The main characteristics of the offset smoker are that the cooking chamber is usually cylindrical in shape, with a shorter, smaller diameter cylinder attached to the bottom of one end for a firebox. To cook the meat, a small fire is lit in the firebox, where airflow is tightly controlled. The heat and smoke from the fire is drawn through a connecting pipe or opening into the cooking chamber. The heat and smoke cook and flavor the meat before escaping through an exhaust vent at the opposite end of the cooking chamber. Most manufacturers' models are based on this simple but effective design, and this is what most people picture when they think of a "BBQ smoker." Even large capacity commercial units use this same basic design of a separate, smaller fire box and a larger cooking chamber.

Upright drum smoker

The upright drum smoker (also referred to as an ugly drum smoker or UDS) is exactly what its name suggests; an upright steel drum that has been modified for the purpose of pseudo-indirect hot smoking. There are many ways to accomplish this, but the basics include the use of a complete steel drum, a basket to hold charcoal near the bottom, and cooking rack (or racks) near the top; all covered by a vented lid of some sort. They have been built using many different sizes of steel drums (30 US gallons (110 l; 25 imp gal), 55 US gallons (210 l; 46 imp gal), and 85 US gallons (320 l; 71 imp gal) for example), but the most popular size is the common 55 gallon drum.

This design is similar to smoking with indirect heat due to the distance from the coals and the racks (typically 24 inches (61 cm)). The temperatures used for smoking are controlled by limiting the amount of air intake at the bottom of the drum, and allowing a similar amount of exhaust out of vents in the lid. UDSs are very efficient with fuel consumption and flexible in their abilities to produce proper smoking conditions, with or without the use of a water pan or drip pan. Most UDS builders/users would say a water pan defeats the true pit BBQ nature of the UDS, as the drippings from the smoked meat should land on the coals, burning up, and imparting a unique flavor one cannot get with a water pan.

Vertical water smoker

A vertical water smoker (also referred to as a bullet smoker because of its shape) is a variation of the upright drum smoker. It uses charcoal or wood to generate smoke and heat, and contains a water bowl between the fire and the cooking grates. The water bowl serves to hold the temperature down and also to add humidity to the smoke chamber. In addition, the bowl catches any drippings from the meat that may cause a flare up. Vertical water smokers are extremely temperature stable and require very little adjustment once the desired temperature has been reached. Because of their relatively low cost and stable temperature, they are sometimes used in barbecue competitions where propane and electric smokers are not allowed.

Propane smoker

A propane smoker is designed to allow the smoking of meat in a somewhat more controlled environment. The primary differences are the sources of heat and of the smoke. In a propane smoker, the heat is generated by a gas burner directly under a steel or iron box containing the wood or charcoal that provides the smoke. The steel box has few vent holes, on the top of the box only. By starving the heated wood of oxygen, it smokes instead of burning. Any combination of woods and charcoal may used.

Smoke box method

This more traditional method uses a two box system: The fire box and the food box. The fire box is typically adjacent or under the cooking box, and can be controlled to a finer degree. The heat and smoke from the fire box exhausts into the food box, where it is used to cook and smoke the meat. These may be as simple as an electric heating element with a pan of wood chips placed on it, although more advanced models have finer temperature controls.

Commercial smoke house

Commercial smokehouses, mostly made from stainless steel, have independent systems for smoke generation and cooking. Smoke generators use friction, an electric coil or a small flame to ignite sawdust on demand. Heat from steam coils or gas flames is balanced with live steam or water sprays to control the temperature and humidity. Elaborate air handling systems reduce hot or cold spots, to reduce variation in the finished product. Racks on wheels or rails are used to hold the product and facilitate movement.

Health Risks

Evidence suggests that smoked foods may contain carcinogens. The smoking process contaminates food with polycyclic aromatic hydrocarbons (PAHs), which are known carcinogens, so, in theory, consuming smoked food increases the risk of gastrointestinal cancer. Some studies have found a positive statistical correlation between intestinal tract cancer and the frequent intake of smoked foods. In one Hungarian longitudinal study, a district in which home-smoked meat was the predominant protein source consumed showed that the incidence of stomach cancer, relative to all other cancers, was nearly twice as high (47%–50%) as that of the general Hungarian population (29.9%).

Other sources, however, while agreeing that PAHs help cause cancer, note the dearth of research studies proving a strong correlation between the intake of smoked foods and increased cancer risk. The National Cancer Institute emphasizes that "population studies have not established a definitive link between ... cooked meats and cancer in humans," but suggests individuals reduce their exposure to PAHs.

Cooking and Health

When heat is used in the preparation of food, it can kill or inactivate potentially harmful organisms, such as bacteria and viruses, as well as various parasites such as tapeworms and Toxoplasma gondii. Food poisoning and other illness from uncooked or poorly-prepared food may be caused by bacteria such as pathogenic strains of Escherichia coli, Salmonella typhimurium and Campylobacter, viruses such as noroviruses, and protozoa such as Entamoeba histolytica. Parasites may be introduced through salad, meat that is uncooked or done rare, and unboiled water.

The sterilizing effect of cooking will depend on temperature, cooking time, and technique used. However, some bacteria such as Clostridium botulinum or Bacillus cereus, can form spores that survive cooking, which then germinate and regrow after the food has cooled. It is therefore recommended that cooked food should not be reheated more than once to avoid repeated growths that allow the bacteria to proliferate to dangerous level.

Cooking prevents many foodborne illnesses that would otherwise occur if the food was eaten raw. Cooking also increases the digestibility of some foods such as grains. Many foods, when raw, are inedible, and some are poisonous. For example kidney beans are toxic when raw or improperly cooked, due to the presence of phytohaemagglutinin which can be inactivated after cooking for at least ten minutes at 100 °C. Slow cooker however may not reach the desired temperature and cases of poisoning from red beans cooked in slow cooker have been reported.

Preparation, handling, and storage of food are other considerations in food safety. The temperature range from 41°F to 135 °F is the "Danger zone" where bacteria is likely to proliferate, food therefore should not be stored in this temperature range. Washing of hands and surfaces, and avoidance of cross-contamination are good practices in food safety. Food prepared on plastic cutting boards may be less likely to harbor bacteria than wooden ones, other research however suggested otherwise. Washing and sanitizing cutting boards is highly recommended, especially after use with raw meat, poultry, or seafood. Hot water and soap followed by a rinse with an diluted antibacterial cleaner, or a trip through a dishwasher with a "sanitize" cycle, are effective methods for reducing the risk of illness due to contaminated cooking implements.

Effects on Nutritional Content of Food

Proponents of Raw foodism argue that cooking food increases the risk of some of the detrimental effects on food or health. They point out that the cooking of vegetables and fruit containing vitamin C both elutes the vitamin into the cooking water and degrades the vitamin through oxidation. Peeling vegetables can also substantially reduce the vitamin C content, especially in the case of potatoes where most vitamin C is in the skin. However, research using an artificial gut has shown that in the specific case of carotenoids a greater proportion is absorbed from cooked vegetables than from raw vegetables.

German research in 2003 showed significant benefits in reducing breast cancer risk when large amounts of raw vegetable matter are included in the diet. The authors attribute some of this effect to heat-labile phytonutrients. Sulforaphane, which may be found in vegetables such as broccoli, has been shown to be protective against prostate cancer, however, much of it is destroyed when the vegetable is boiled.

Cooking and Carcinogens

In a human epidemiological analysis by Richard Doll and Richard Peto in 1981, diet was estimated to cause perhaps around 35% of cancers. Some of these cancers may be caused by carcinogens in food generated during cooking process, although it is often difficult to identify the specific components in diet that serve to increase cancer risk. Many food, such as beef steak and broccoli, contain low concentrations of both carcinogens and anticarcinogens.

Several studies published since 1990 indicate that cooking meat at high temperature creates heterocyclic amines (HCAs), which are thought to increase cancer risk in humans. Researchers at the National Cancer Institute found that human subjects who ate beef rare or medium-rare had less than one third the risk of stomach cancer than those who ate beef medium-well or well-done. While eating meat raw may be the only way to avoid HCAs fully, the National Cancer Institute states that cooking meat below 212 °F (100 °C) creates "negligible amounts" of HCAs. Also, microwaving meat before cooking may reduce HCAs by 90%. Nitrosamines, present in processed and cooked foods, have also been noted as being carcinogenic, being linked to colon cancer.

Research has shown that grilling, barbecuing and smoking meat and fish increases levels of carcinogenic Polycyclic aromatic hydrocarbons (PAH). In Europe, grilled meat and smoked fish generally only contribute a small proportion of dietary PAH intake since they are a minor component of diet – most intake comes from cereals, oils and fats. However, in the US, grilled/barbecued meat is the second highest contributor of the mean daily intake of benzo[a]pyrene at 21% after 'bread, cereal and grain' at 29%.

Baking, grilling or broiling food, especially starchy foods, until a toasted crust is formed generates significant concentrations of acrylamide, a possible carcinogen.

Other Health Issues

Cooking dairy products may reduce a protective effect against colon cancer. Researchers at the University of Toronto suggest that ingesting uncooked or unpasteurized dairy products may reduce the risk of colorectal cancer. Mice and rats fed uncooked sucrose, casein, and beef tallow had one-third to one-fifth the incidence of microadenomas as the mice and rats fed the same ingredients cooked. This claim, however, is contentious. According to the Food and Drug Administration of the United States, health benefits claimed by raw milk advocates do not exist. "The small quantities of antibodies in milk are not absorbed in the human intestinal tract," says Barbara Ingham, PhD, associate professor and extension food scientist at the University of Wisconsin-Madison. "There is no scientific evidence that raw milk contains an anti-arthritis factor or that it enhances resistance to other diseases."

Heating sugars with proteins or fats can produce Advanced glycation end products. These have been linked to ageing and health conditions such as diabetes.Deep fried food in restaurants may contain high level of trans fat which is known to increase level of low-density lipoprotein that may increase risk of heart diseases and other conditions. However, many fast food chains have now switched to trans-fat-free alternatives for deep-frying.

References

Carcinogens and Anticarcinogens in the Human Diet. National Academy Press. (1996).

Christine Dell'Amore in Chicago. (2009). "Cooking Gave Humans Edge Over Apes?". National Geographic News.

Corpet DE, Yin, Zhang XM, *et al.* (1990). "Colonic protein fermentation and promotion of colon carcinogenesis by thermolyzed casein". *Nutr Cancer* 23 (3): 271–81.

Gorman, RM.(2008). "Cooking up bigger brains". *Scientific American* 298 (1): 102, 104–5.doi:10.1038/scientificamerican0108-102.

Jin, Y.; Wang, M.; Rosen, R. T.; Ho, C. T. (1999). "Thermal Degradation of Sulforaphane in Aqueous Solution". *Journal of Agricultural and Food Chemistry* 47 (8): 3121–3123.

6

Methods of Food Preservation

Food preservation is the process of treating and handling food to stop or slow down food spoilage, loss of quality, edibility or nutritional value and thus allow for longer food storage.

Preservation usually involves preventing the growth of bacteria, fungi (such as yeasts), and other micro-organisms (although some methods work by introducing benign bacteria, or fungi to the food), as well as retarding the oxidation of fats which cause rancidity. Food preservation can also include processes which inhibit visual deterioration, such as the enzymatic browning reaction in apples after they are cut, which can occur during food preparation.

Many processes designed to preserve food will involve a number of food preservation methods. Preserving fruit by turning it into jam, for example, involves boiling (to reduce the fruit's moisture content and to kill bacteria, yeasts, etc.), sugaring (to prevent their re-growth) and sealing within an airtight jar (to prevent recontamination). There are many traditional methods of preserving food that limit the energy inputs and reduce carbon footprint.

Maintaining or creating nutritional value, texture and flavour is an important aspect of food preservation, although, historically, some methods drastically altered the character of the food being preserved. In many cases these changes have now come to be seen as desirable qualities – cheese, yoghurt and pickled onions being common examples.

Refrigeration

Refrigeration preserves food by slowing down the growth and reproduction of micro-organisms and the action of enzymes which cause food to rot. The introduction of commercial and domestic refrigerators drastically improved the diets of many in the Western world by allowing foods such as fresh fruit, salads and dairy products to be stored safely for longer periods, particularly during warm weather.

Freezing

Freezing is also one of the most commonly used processes commercially and domestically for preserving a very wide range of food including prepared food stuffs which would not have required freezing in their unprepared state. For example, potato waffles are stored in the freezer, but potatoes themselves require only a cool dark place to ensure many months' storage. Cold stores provide large volume, long-term storage for strategic food stocks held in case of national emergency in many countries.

Vacuum packing

Vacuum-packing stores food in a vacuum environment, usually in an air-tight bag or bottle. The vacuum environment strips bacteria of oxygen needed for survival, slowing spoiling. Vacuum-packing is commonly used for storing nuts to reduce loss of flavor from oxidation.

Salting

Salting or curing draws moisture from the meat through a process of osmosis. Meat is cured with salt or sugar, or a combination of the two. Nitrates and nitrites are also often used to cure meat and contribute the characteristic pink color, as well as inhibition of Clostridium botulinum.

Sugaring

Sugar is used to preserve fruits, either in syrup with fruit such as apples, pears, peaches, apricots, plums or in crystallized form where the preserved material is cooked in sugar to the point of crystallisation and the resultant product is then stored dry. This method is used for the skins of citrus fruit (candied peel), angelica and ginger. A modification of this process produces glacé fruit such as glacé cherries where the fruit is preserved in sugar but is then extracted from the syrup and sold, the preservation being maintained by the sugar content of the fruit and the superficial coating of syrup. The use of sugar is often combined with alcohol for preservation of luxury products such as fruit in brandy or other spirits. These should not be confused with fruit flavored spirits such as cherry brandy or Sloe gin.

Smoking

moking is used to lengthen the shelf life of perishable food items. This effect is achieved by exposing the food to smoke from burning plant materials such as wood. Most commonly subjected to this method of food preservation are meats and fish that have undergone curing. Fruits and vegetables like paprika, cheeses, spices, and ingredients for making drinks such as malt and tea leaves are also smoked, but mainly for cooking or flavoring them. It is one of the oldest food preservation methods, which probably arose after the development of cooking with fire.

Artificial Food Additives

Preservative food additives can be antimicrobial; which inhibit the growth of bacteria or fungi, including mold, or antioxidant; such as oxygen absorbers, which inhibit the oxidation of food constituents. Common antimicrobial preservatives include calcium propionate, sodium nitrate, sodium nitrite, sulfites (sulfur dioxide, sodium bisulfite, potassium hydrogen sulfite, etc.) and disodium EDTA. Antioxidants include BHA and BHT. Other preservatives include formaldehyde (usually in solution), glutaraldehyde (kills insects), ethanol and methylchloroisothiazolinone.

Pickling

Pickling is a method of preserving food in an edible anti-microbial liquid. Pickling can be broadly categorized into two categories: chemical pickling and fermentation pickling.

In chemical pickling, the food is placed in an edible liquid that inhibits or kills bacteria and other micro-organisms. Typical pickling agents include brine (high in salt), vinegar, alcohol, and vegetable oil, especially olive oil but also many other oils. Many chemical pickling processes also involve heating or boiling so that the food being preserved becomes saturated with the pickling agent. Common chemically pickled foods include cucumbers, peppers, corned beef, herring, and eggs, as well as mixed vegetables such as piccalilli.

In fermentation pickling, the food itself produces the preservation agent, typically by a process that produces lactic acid. Fermented pickles include sauerkraut, nukazuke, kimchi, surströmming, and curtido. Some pickled cucumbers are also fermented.

Lye

Sodium hydroxide (lye) makes food too alkaline for bacterial growth. Lye will saponify fats in the food, which will change its flavor and texture. Lutefisk uses lye in its preparation, as do some olive recipes. Modern recipes for century eggs also call for lye. Masa harina and hominy use agricultural lime in their preparation and this is often misheard as 'lye'.

Canning and Bottling

Canning involves cooking food, sealing it in sterile cans or jars, and boiling the containers to kill or weaken any remaining bacteria as a form of sterilization. It was invented by Nicolas Appert. Foods have varying degrees of natural protection against spoilage and may require that the final step occur in a pressure cooker. High-acid fruits like strawberries require no preservatives to can and only a short boiling cycle, whereas marginal fruits such as tomatoes require longer boiling and addition of other acidic elements. Low acid foods, such as vegetables and meats require pressure canning. Food preserved by canning or bottling is at immediate risk of spoilage once the can or bottle has been opened.

Lack of quality control in the canning process may allow ingress of water or micro-organisms. Most such failures are rapidly detected as decomposition within the can causes gas production and the can will swell or burst. However, there have been examples of poor manufacture and poor hygiene allowing contamination of canned food by the obligate anaerobe Clostridium botulinum, which produces an acute toxin within the food, leading to severe illness or death. This organism produces no gas or obvious taste and remains undetected by taste or smell. Its toxin is denatured by cooking, though. Cooked mushrooms, handled poorly and then canned, can support the growth of Staphylococcus aureus, which produces a toxin that is not destroyed by canning or subsequent reheating.

Jellying

Food may be preserved by cooking in a material that solidifies to form a gel. Such materials include gelatine, agar, maize flour and arrowroot flour. Some foods naturally form a protein gel when cooked such as eels and elvers, and sipunculid worms which are a delicacy in Xiamen in Fujian province of the People's Republic of China. Jellied eels are a delicacy in the East End of London where they are eaten with mashed potatoes. Potted meats in aspic, were a common way of serving meat off-cuts in the UK until the 1950s. Many jugged meats are also jellied.

Jugging

Meat can be preserved by jugging, the process of stewing the meat (commonly game or fish) in a covered earthenware jug or casserole. The animal to be jugged is usually cut into pieces, placed into a tightly-sealed jug with brine or gravy, and stewed. Red wine and/or the animal's own blood is sometimes added to the cooking liquid. Jugging was a popular method of preserving meat up until the middle of the 20th century.

Irradiation

Irradiation of food is the exposure of food to ionizing radiation; either high-energy electrons or X-rays from accelerators, or by gamma rays (emitted from radioactive sources as Cobalt-60 or Caesium-137). The treatment has a range of effects, including killing bacteria, molds and insect pests, reducing the ripening and spoiling of fruits, and at higher doses inducing sterility. The technology may be compared to pasteurization; it is sometimes called 'cold pasteurization', as the product is not heated. Irradiation is not effective against viruses or prions, it cannot eliminate toxins already formed by microorganisms.

The radiation process is unrelated to nuclear energy, but it may use the radiation emitted from radioactive nuclides produced in nuclear reactors. Ionizing radiation is hazardous to life (hence its usefulness in sterilisation); for this reason irradiation facilities have a heavily shielded irradiation room where the process takes place. Radiation safety procedures ensure that neither the workers in such facility nor the environment receive any radiation dose from the facility. Irradiated food does not become radioactive, and national and international expert bodies have declared food irradiation as wholesome. However, the wholesomeness of consuming such food is disputed by opponents and consumer organizations. National and international expert bodies have declared food irradiation as 'wholesome';

UN-organizations as WHO and FAO are endorsing to use food irradiation. International legislation on whether food may be irradiated or not varies worldwide from no regulation to full banning. Irradiation may allow lower quality or contaminated foodstuffs to be rendered marketable.

It is estimated that about 500,000 tons of food items are irradiated per year worldwide in over 40 countries. These are mainly spices and condiments with an increasing segment of fresh fruit irradiated for fruit fly quarantine. food preservation is very good process

Pulsed Electric Field Processing

Pulsed electric field (PEF) processing is a method for processing cells by means of brief pulses of a strong electric field. PEF holds potential as a type of low temperature alternative pasteurization process for sterilizing food products. In PEF processing, a substance is placed between two electrodes, then the pulsed electric field is applied. The electric field enlarges the pores of the cell membranes which kills the cells and releases their contents. PEF for food processing is a developing technology still being researched. There have been limited industrial applications of PEF processing for the pasteurization of fruit juices.

Modified Atmosphere

Modifying atmosphere is a way to preserve food by operating on the atmosphere around it. Salad crops which are notoriously difficult to preserve are now being packaged in sealed bags with an atmosphere modified to reduce the oxygen (O_2) concentration and increase the carbon dioxide (CO_2) concentration. There is concern that although salad vegetables retain their appearance and texture in such conditions, this method of preservation may not retain nutrients, especially vitamins. Grains may be preserved using carbon dioxide by one of two methods; either using a block of dry ice placed in the bottom and the can is filled with grain or the container can be purged from the bottom by gaseous carbon dioxide from a cylinder or bulk supply vessel.

Carbon dioxide prevents insects, and depending on concentration, mold, and oxidation from damaging the grain. Grain stored in this way can remain edible for five years.

Nitrogen gas (N_2) at concentrations of 98% or higher is also used effectively to kill insects in grain through hypoxia. However, carbon dioxide has an advantage in this respect as it kills organisms through hypercarbia and depending on concentration hypoxia and, requiring concentrations of above 35%, or so. This makes carbon dioxide preferable for fumigation in situations where a hermetic seal cannot be maintained.

Air-tight storage of grains (sometimes called hermetic storage) relies on the respiration of grain, insects and fungi which can modify the enclosed atmosphere sufficiently to control insect pests. This is a method of great antiquity, as well as having modern equivalents. The success of the method relies on have the correct mix of sealing, grain moisture and temperature.

High Pressure Food Preservation

High pressure food preservation refers to high pressure used for food preservation. "Pressed inside a vessel exerting 70,000 pounds per square inch (480 MPa) or more, food can be processed so that it retains its fresh appearance, flavour, texture and nutrients while disabling harmful microorganisms and slowing spoilage. By 2005 the process was being used for products ranging from orange juice to guacamole to deli meats and widely sold.

Burial in the Ground

Burial of food can preserve it due to a variety of factors: lack of light, lack of oxygen, cool temperatures, pH level, or desiccants in the soil. Burial may be combined with other methods such as salting or fermentation. Most foods can be preserved in soil that is very dry and salty (thus a desiccant), or soil that is frozen.

Many root vegetables are very resistant to spoilage and require no other preservation than storage in cool dark conditions, for example by burial in the ground, such as in a storage clamp. Century eggs are created by placing eggs in alkaline mud (or other alkaline substance) resulting in their "inorganic" fermentation through raised pH instead of spoiling. The fermentation preserves them and breaks down some of the complex, less flavorful proteins and fats into simpler more flavorful ones. Cabbage was traditionally buried in the fall in northern farms in the USA for preservation. Some methods keep it crispy while other methods produce sauerkraut[citation needed]. A similar process is used in the traditional production of kimchi. Sometimes meat is buried under conditions which cause preservation. If buried on hot coals or ashes, the heat can kill pathogens, the dry ash can desiccate, and the earth can block oxygen and further contamination. If buried where the earth is very cold, the earth acts like a refrigerator.

Controlled Use of Micro-organism

Some foods, such as many cheeses, wines, and beers will keep for a long time because their production uses specific micro-organisms that combat spoilage from other less benign organisms. These micro-organisms keep pathogens in check by creating an environment toxic for themselves and other micro-organisms by producing acid or alcohol. Starter micro-organisms, salt, hops, controlled (usually cool) temperatures, controlled (usually low) levels of oxygen and/or other methods are used to create the specific controlled conditions that will support the desirable organisms that produce food fit for human consumption.

Biopreservation

Biopreservation is the use of natural or controlled microbiota or antimicrobials as a way of preserving food and extending its shelf life. Beneficial bacteria or the fermentation products produced by these bacteria are used in biopreservation to control spoilage and render pathogens inactive in food. It is a benign ecological approach which is gaining increasing attention.

Of special interest are lactic acid bacteria (LAB). Lactic acid bacteria have antagonistic properties which make them particularly useful as biopreservatives. When LABs compete for nutrients, their metabolites often include active antimicrobials such as lactic and acetic acid, hydrogen peroxide, and peptide bacteriocins. Some LABs produce the antimicrobial nisin which is a particularly effective preservative.

These days LAB bacteriocins are used as an integral part of hurdle technology. Using them in combination with other preservative techniques can effectively control spoilage bacteria and other pathogens, and can inhibiting the activities of a wide spectrum of organisms, including inherently resistant Gram-negative bacteria.

Hurdle Technology

Hurdle technology is a method of ensuring that pathogens in food products can be eliminated or controlled by combining more than one approach. These approaches can be thought of as "hurdles" the pathogen has to overcome if it is to remain active in the food. The right combination of hurdles can ensure all pathogens are eliminated or rendered harmless in the final product.

Hurdle technology has been defined by Leistner as an intelligent combination of hurdles which secures the microbial safety and stability as well as the organoleptic and nutritional quality and the economic viability of food products. The organoleptic quality of the food refers to its sensory properties, that is its look, taste, smell and texture.

Examples of hurdles in a food system are high temperature during processing, low temperature during storage, increasing the acidity, lowering the water activity or redox potential, or the presence of preservatives or biopreservatives. According to the type of pathogens and how risky they are, the intensity of the hurdles can be adjusted individually to meet consumer preferences in an economical way, without sacrificing the safety of the product.

References

Alasalvar C (2010) *Seafood Quality, Safety and Health Applications* John Wiley and Sons, Page 203.

Alzamora SM, Tapia MS and López-Malo A (2000) *Minimally processed fruits and vegetables: fundamental aspects and applications* Springer.

FAO: Preservation techniques Fisheries and aquaculture department, Rome. Updated 27 May 2005. Retrieved 14 March 2011.

Lee S (2004) "Microbial Safety of Pickled Fruits and Vegetables and Hurdle Technology" *Internet Journal of Food Safety*, 4: 21-32.

Leistner L (1995) "Principles and applications of hurdle technology" In Gould GW (Ed.) *New Methods of Food Preservation*, Springer, pp. 1-21.

Yousef AE and Carolyn Carlstrom C (2003) *Food microbiology: a laboratory manual* Wiley.

7

Processing and Preservation of Fruits and Vegetables

The main objective of fruit and vegetable processing is to supply wholesome, safe, nutritious and acceptable food to consumers throughout the year. Fruit and vegetable processing projects also aim to replace imported products like squash, yams, tomato sauces, pickles, etc., besides earning foreign exchange by exporting finished or semi-processed products. The fruit and vegetable processing activities have been set up, or have to be established in developing countries for one or other of the following reasons:

— diversification of the economy, in order to reduce present dependence on one export commodity;
— government industrialisation policy;
— reduction of imports and meeting export demands;
— stimulate agricultural production by obtaining marketable products;
— generate both rural and urban employment;
— reduce fruit and vegetable losses;
— improve farmers' nutrition by allowing them to consume their own processed fruit and vegetables during the off-season;
— generate new sources of income for farmers/artisans;
— develop new value-added products.

Practically any fruit and vegetable can be processed, but some important factors which determine whether it is worthwhile are:

a) the demand for a particular fruit or vegetable in the processed form;

b) the quality of the raw material, *ie*. whether it can withstand processing;

c) regular supplies of the raw material.

For example, a particular variety of fruit which may be excellent to eat fresh is not necessarily good for processing. Processing requires frequent handling, high temperature and pressure. Many of the ordinary table varieties of tomatoes, for instance, are not suitable for making paste or other processed products. A particular mango or pineapple may be very tasty eaten fresh, but when it goes to the processing centre it may fail to stand up to the processing requirements due to variations in its quality, size, maturity, variety and so on.

Even when a variety can be processed, it is not suitable unless large and regular supplies are made available. An important processing centre or a factory cannot be planned just to rely on seasonal gluts; although it can take care of the gluts it will not run economically unless regular supplies are guaranteed. To operate a fruit and vegetable processing centre efficiently it is of utmost importance to pre-organise growth, collection and transport of suitable raw material, either on the nucleus farm basis or using outgrowers.

PLANNING

The secret of a well planned fruit and vegetable processing centre is that it must be designed to operate for as many months of the year as possible. This means the facilities, the buildings, the material handling and the equipment itself must be inter-linked and coordinated properly to allow as many products as possible to be handled at the same time, and yet the equipment must be versatile enough to be able to handle many products without major alterations.

A typical processing centre or factory should process four or five types of fruits harvested at different times of the year and two or three vegetables. This processing unit must also be capable of handling dried/dehydrated finished products, juices, pickles, tomato juice, ketchup and paste, jams, jellies and marmalades, semi-processed fruit products. Advanced planning is necessary to process a large range of products in varied weather and temperature conditions, each requiring a special set of manufacturing and packaging formulae. The end result of the efforts should be a well-managed processing unit with lower initial investment.A unit which is sensibly laid out and where one requirement co-relates to another, with a sound costing analysis, leads to an integrated operation. Instead of over-sophisticated machinery, a sensible simple processing unit may be required when planned production is not very large and is geared mainly to meet the demand of the domestic market.

The basic objective is to choose the location which minimises the average production cost, including transport and handling. It is an advantage, all other things being equal, to locate a processing unit near the fresh raw material supply. It is a necessity for proper handling of the perishable raw materials, it allows the processing unit to allow the product to reach its best stage of maturation and lessens injury from handling and deterioration from changes during long transportation after harvesting. An adequate supply of good water, availability of manpower, proximity to rail or road transport facilities and adequate markets are other important requirements.

Processing Systems

a) *Small-Scale Processing*: This is done by small-scale farmers for personal subsistence or for sale in nearby markets. In this system, processing requires little investment: however, it is time consuming and tedious. Until recently, small-scale processing satisfied the needs of rural and urban populations. However, with the rising rates of population and urbanisation growth and their more diversified food demands, there is need for more processed and diversified types of food.

b) *Intermediate-Scale Processing*: In this scale of processing, a group of small-scale processors pool their resources. This can also be done by individuals. Processing is based on the technology used by small-scale processors with differences in the type and capacity of equipment used. The raw materials are usually grown by the processors themselves or are purchased on contract from other farmers. These operations are usually located on the production site of in order to assure raw materials availability and reduce cost of transport. This system of processing can provide quantities of processed products to urban areas.

c) *Large-Scale Processing*: Processing in this system is highly mechanised and requires a substantial supply of raw materials for economical operation. This system requires a large capital investment and high technical and managerial skills. Because of the high demand for foods in recent years many large-scale factories were established in developing countries. Some succeeded, but the majority failed, especially in West Africa. Most of the failures were related to high labour inputs and relatively high cost, lack of managerial skills, high cost and supply instability of raw materials and changing governmental policies. Perhaps the most important reason for failure was lack of adequate quantity and regularity of raw material supply to factories. Despite the failure of these commercial operations, they should be able to succeed with better planning and management, along with the undertaking of more in-depth feasibility studies.

It can be concluded that all three types of processing systems have a place in developing countries to complement crop production to meet food demand. Historically, however, small and intermediate scale processing proved to be more successful than large-scale processing in developing countries.

Processing Technologies

The basis for choosing a processing technology for developing countries ought to be to combine labour, material resources and capital so that not only the type and quantity of goods and services produced are taken into account, but also the distribution of their benefits and the prospects of overall growth. These should include:

a) increasing farmer/artisan income by the full utilisation of available indigenous raw material and local manufacturing of part or all processing equipment;

b) cutting production costs by better utilisation of local natural resources (solar energy) and reducing transport costs;

c) generating and distributing income by decentralising processing activities and involving different beneficiaries in processing activities (investors, newly employed, farmers and small-scale industry);

d) maximising national output by reducing capital expenditure and royalty payments, more effectively developing balance-of-payments deficits through minimising imports (equipment, packing material, additives), and maximising export-oriented production;

e) maximising availability of consumer goods by maximisation of high-quality, standard processed produce for internal and export markets, reducing post-harvest losses, giving added value to indigenous crops and increasing the volume and quality of agricultural output.

Knowledge and control of the means of production, local manufacturing of processing equipment and development of appropriate/new technologies and more suitable raw material for processing must all be better researched.

Decentralisation of activities must be maintained and coordinated. The introduction of more sophisticated processing equipment and packaging material must be subordinated to internal and export marketing references. Choosing a technology solely to maximise profits can actually work against true development. Choice should also be based on a solid, long-term market opportunity to ensure viability. The internal market should be given greater consideration, safeguarded and supported. Training courses, at all levels, in processing and preservation of indigenous crops, must be expanded.

Properties of Fruit and Vegetables

Fruit and vegetables have many similarities with respect to their compositions, methods of cultivation and harvesting, storage properties and processing. In fact, many vegetables may be considered fruit in the true botanical sense. Botanically, fruits are those portions of the plant which house seeds. Therefore such items as tomatoes, cucumbers, eggplant, peppers, and others would be classified as fruits on this basis.However, the important distinction between fruit and vegetables has come to be made on an usage basis. Those plant items that are generally eaten with the main course of a meal are considered to be vegetables. Those that are commonly eaten as dessert are considered fruits. Vegetables are derived from various parts of plants and it is sometimes useful to associate different vegetables with the parts of the plant they represent since this provides clues to some of the characteristics we may expect in these items. A classification of vegetables based on morphological features is seen in Table 1.

Fruit as a dessert item, is the mature ovaries of plants with their seeds. The edible portion of most fruit is the fleshy part of the pericarp or vessel surrounding the seeds. Fruit in general is acidic and sugary. They commonly are grouped into several major divisions, depending principally upon botanical structure, chemical composition and climatic requirements. Berries are fruit which are generally small and quite fragile. Grapes are also physically fragile and grow in clusters. Melons, on the other hand, are large and have a tough outer rind. Drupes (stone fruit) contain single pits and include such items as apricots, cherries, peaches and plums. Pomes contain many pits, and are represented by apples, quince and pears.

Table 1: Classification of Vegetables

Category	*Examples*
Earth vegetables roots	sweet potatoes, carrots
modified stems tubers	potatoes
modified buds bulbs	onions, garlic
leaves	cabbage, spinach, lettuce
petioles (leaf stalk)	celery, rhubarb
flower buds	cauliflower, artichokes
sprouts, shoots (young stems)	asparagus, bamboo shoots
legumes	peas, green beans
cereals	sweet corn
vine fruits	squash, cucumber
berry fruits	tomato, egg plant
tree fruits	avocado, breadfruit

Citrus fruit like oranges, grapefruit and lemons are high in citric acid. Tropical and subtropical fruits include bananas, dates, figs, pineapples, mangoes, and others which require warm climates, but exclude the separate group of citrus fruits. The compositions of representative vegetables and fruits in comparison with a few of the cereal grains are seen in Table 2.

Compositions of vegetables and fruit not only vary for a given kind in according to botanical variety, cultivation practices, and weather, but change with the degree of maturity prior to harvest, and the condition of ripeness, which is progressive after harvest and is further influenced by storage conditions. Nevertheless, some generalisations can be made. Most fresh vegetables and fruit are high in water content, low in protein, and low in fat. In these cases water contents will generally be greater than 70% and frequently greater than 85%. Commonly protein content will not be greater than 3.5% or fat content greater than 0.5 %. Exceptions exist in the case of dates and raisins which are substantially lower in moisture but cannot be considered fresh in the same sense as other fruit. Legumes such as peas and certain beans are higher in protein; a few vegetables such as sweet corn which are slightly higher in fat and avocados which are substantially higher in fat.

Table 2: Typical percentage composition of foods from plant origin

Food	*Carbohydrate*	*Protein*	*Fat*	*Ash*	*Water*
Cereals					
wheat flour, white	73.9	10.5	1.9	1.7	12
rice, milled, white	78.9	6.7	0.7	0.7	13
maize, whole grain	72.9	9.5	4.3	1.3	12
Earth vegetables					
potatoes, white	18.9	2.0	0.1	1.0	78
sweet potatoes	27.3	1.3	0.4	1.0	70
Vegetables					
carrots	9.1	1.1	0.2	1.0	88.6
radishes	4.2	1.1	0.1	0.9	93.7
asparagus	4.1	2.1	0.2	0.7	92.9
beans, snap, green	7.6	2.4	0.2	0.7	89.1
peas, fresh	17.0	6.7	0.4	0.9	75.0
lettuce	2.8	1.3	0.2	0.9	94.8
Fruit					
banana	24.0	1.3	0.4	0.8	73.5
orange	11.3	0.9	0.2	0.5	87.1
apple	15.0	0.3	0.4	0.3	84.0
strawberries	8.3	0.8	0.5	0.5	89.9

Vegetables and fruit are important sources of both digestible and indigestible carbohydrates. The digestible carbohydrates are present largely in the form of sugars and starches while indigestible cellulose provides roughage which is important to normal digestion. Fruit and vegetables are also important sources of minerals and certain vitamins, especially vitamins A and C. The precursors of vitamin A, including beta-carotene and certain other carotenoids, are to be found particularly in the yellow-orange fruit and vegetables and in the green leafy vegetables. Citrus fruit are excellent sources of vitamin C, as are green leafy vegetables and tomatoes.

Deterioration Reactions

Overall deterioration reactions in fruits and vegetables are presented below.

Changes in Enzymes

Enzymes which are endogenous to plant tissues can have undesirable or desirable consequences. Examples involving endogenous enzymes include

a) the post-harvest senescence and spoilage of fruit and vegetables;
b) oxidation of phenolic substances in plant tissues by phenolase (leading to browning);
c) sugar—starch conversion in plant tissues by amylases;
d) post-harvest demethylation of pectic substances in plant tissues (leading to softening of plant tissues during ripening, and firming of plant tissues during processing).

The major factors useful in controlling enzyme activity are: temperature, water activity, pH, chemicals which can inhibit enzyme action, alteration of substrates, alteration of products and pre-processing control.

Chemical Changes

Sensory quality

The two major chemical changes which occur during the processing and storage of foods and lead to a deterioration in sensory quality are lipid oxidation and non-enzymatic browning. Chemical reactions are also responsible for changes in the colour and flavour of foods during processing and storage. Lipid oxidation rate and course of reaction is influenced by light, local oxygen concentration, high temperature, the presence of catalysts (generally transition metals such as iron and copper) and water activity. Control of these factors can significantly reduce the extent of lipid oxidation in foods.

Non-enzymic browning is one of the major causes of deterioration which occurs during storage of dried and concentrated foods. The non-enzymic browning, or Maillard reaction, can be divided into three stages: a) early Maillard reactions which are chemically well-defined steps without browning; b) advanced Maillard reactions which lead to the formation of volatile or soluble substances; and c) final Maillard reactions leading to insoluble brown polymers.

Colour changes

Chlorophylls: Almost any type of food processing or storage causes some deterioration of the chlorophyll pigments. Phenophytinisation (with consequent formation of a dull olivebrown phenophytin) is the major change; this reaction is accelerated by heat and is acid catalysed. Other reactions are also possible. For example, dehydrated products such as green peas and beans packed in clear glass containers undergo photo-oxidation and loss of desirable colour.

Anthocyanins: These are a group of more than 150 reddish water-soluble pigments that are very widespread in the plant kingdom. The rate of anthocyanin destruction is pH dependent, being greater at higher pH values. Of interest from a packaging point of view is the ability of some anthocyanins to form complexes with metals such as Al, Fe, Cu and Sn. These complexes generally result in a change in the colour of the pigment (for example, red sour cherries react with tin to form a purple complex) and are therefore undesirable. Since metal packaging materials such as cans could be sources of these metals, they are usually coated with special organic linings to avoid these undesirable reactions.

Carotenoids: The carotenoids are a group of mainly lipid soluble compounds responsible for many of the yellow and red colours of plant and animal products. The main cause of carotenoid degradation in foods is oxidation. The mechanism of oxidation in processed foods is complex and depends on many factors. The pigments may auto-oxidise by reaction with atmospheric oxygen at rates dependent on light, heat and the presence of pro- and antioxidants.

Flavour changes

In fruit and vegetables, enzymically generated compounds derived from long-chain fatty acids play an extremely important role in the formation of characteristic flavours. In addition, these types of reactions can lead to significant off-flavours. Enzyme-induced oxidative breakdown of unsaturated fatty acids occurs extensively in plant tissues and this yield characteristic aromas associated with some ripening fruits and disrupted tissues. The permeability of packaging materials is of importance in retaining desirable volatile components within packages, or in permitting undesirable components to permeate through the package from the ambient atmosphere.

Nutritional quality changes

The four major factors which affect nutrient degradation and can be controlled to varying extents by packaging are light, oxygen concentration, temperature and water activity. However, because of the diverse nature of the various nutrients as well as the chemical heterogeneity within each class of compounds and the complex interactions of the above variables, generalisations about nutrient degradation in foods will inevitably be broad ones.

Vitamins: Ascorbic acid is the most sensitive vitamin in foods, its stability varying markedly as a function of environmental conditions such as pH and the concentration of trace metal ions and oxygen. The nature of the packaging material can significantly affect the stability of ascorbic acid in foods. The effectiveness of the material as a barrier to moisture and oxygen as well as the chemical nature of the surface exposed to the food are important factors. For example, problems of ascorbic acid instability in aseptically packaged fruit juices have been encountered because of oxygen permeability of the package and the oxygen dependence of the ascorbic acid degradation reaction.

Also, because of the preferential oxidation of metallic tin, citrus juices packaged in cans with a tin contact surface exhibit greater stability of ascorbic acid than those in enamelled cans or glass containers. The aerobic and anaerobic degradation reactions of ascorbic acid in reduced-moisture foods have been shown to be highly sensitive to water activity, the reaction rate increasing in an exponential fashion over the water activity range of 0.1-0.8.

Physical Changes

One major undesirable physical change in food powders is the absorption of moisture as a consequence of an inadequate barrier provided by the package; this results in caking. It can occur either as a result of a poor selection of packaging material in the first place, or failure of the package integrity during storage. In general, moisture absorption is associated with increased cohesiveness. Anti-caking agents are very fine powders of an inert chemical substance that are added to powders with much larger particle size in order to inhibit caking and improve flowability.

Studies in onion powders showed that at ambient temperature, caking does not occur at water activities of less than about 0.4. At higher activities, however, ($aw > 0.45$) the observed time to caking is inversely proportional to water activity, and at these levels anti-caking agents are completely ineffective. It appears that while they reduce inter-particle attraction and interfere with the continuity of liquid bridges, they are unable to cover moisture sorption sites.

Biological Changes

Microbiological changes

Micro-organisms can make both desirable and undesirable changes to the quality of foods depending on whether or not they are introduced as an essential part of the food preservation process or arise unintentionally and subsequently grow to produce food spoilage. The two major groups of micro-organisms found in foods are bacteria and fungi, the latter consisting of yeasts and moulds. Bacteria are generally the fastest growing, so that in conditions favourable to both, bacteria will usually outgrow fungi.

Foods are frequently classified on the basis of their stability as non-perishable, semi-perishable and perishable. For example, hermetically sealed and heat processed (*eg*. canned) foods are generally regarded as non-perishable. However, they may become perishable under certain circumstances when an opportunity for recontamination is afforded following processing. Such an opportunity may arise if the can seams are faulty, or if there is excessive corrosion resulting in internal gas formation and eventual bursting of the can.

Spoilage may also take place when the canned food is stored at unusually high temperatures: thermophilic spore-forming bacteria may multiply, causing undesirable changes such as flat sour spoilage. Low moisture content foods such as dried fruit and vegetables are classified as semi-perishable. Frozen foods, though basically perishable, may be classified as semi-perishable provided that they are properly stored at freezer temperatures. The majority of foods (*eg*. meat and fish, milk, eggs and most fresh fruits and vegetables) are classified as perishable unless they have been processed in some way. Often, the only form of processing which such foods receive is to be packaged and kept under controlled temperature conditions.

The species of micro-organisms which cause the spoilage of particular foods are influenced by two factors: a) the nature of the foods and b) their surroundings. These factors are referred to as intrinsic and extrinsic parameters. The intrinsic parameters are an inherent part of the food: pH, a_w, nutrient content, antimicrobial constituents and biological structures. The extrinsic parameters of foods are those properties of the storage environment that affect both the foods and their microorganisms.

The growth rate of the micro-organisms responsible for spoilage primarily depends on these extrinsic parameters: temperature, relative humidity and gas compositions of the surrounding atmosphere. The protection of packaged food from contamination or attack by micro-organisms depends on the mechanical integrity of the package (*eg*. the absence of breaks and seal imperfections), and on the resistance of the package to penetration by micro-organisms. Metal cans which are retorted after filling can leak during cooling, admitting any microorganisms which may be present in the cooling water, even when the double seam is of a high quality. This fact is widely known in the canning industry and is the reason for the mandatory chlorination of cannery cooling water.

Extensive studies on a variety of plastic films and metal foils have shown that microorganisms (including mounds, yeasts and bacteria) cannot penetrate these materials in the absence of pinholes. In practice, however, thin sheets of packaging materials such as aluminium and plastic do contain pinholes. There are several safeguards against the passage of micro-organisms through pinholes in films:

— because of surface tension effects, micro-organisms cannot pass through very small pinholes unless the micro-organisms are suspended in solutions containing wetting agents and the pressure outside the package is greater than that within;

— materials of packaging are generally used in thicknesses such that pinholes are very infrequent and small;

— for applications in which package integrity is essential (such as sterilisation of food in pouches), adequate test methods are available to assure freedom from bacterial recontamination.

Macrobiological changes

Insect pests

Warm humid environments promote insect growth, although most insects will not breed if the temperature exceeds about 35 C° or falls below 10 C°. Also many insects cannot reproduce satisfactorily unless the moisture content of their food is greater than about 11%. The main categories of foods subject to pest attack are cereal grains and products derived from cereal grains, other seeds used as food (especially legumes), dairy products such as cheese and milk powders, dried fruits, dried and smoked meats and nuts.

As well as their possible health significance, the presence of insects and insect excrete in packaged foods may render products unsaleable, causing considerable economic loss, as well as reduction in nutritional quality, production of off-flavours and acceleration of decay processes due to creation of higher temperatures and moisture levels. Early stages of infestation are often difficult to detect; however, infestation can generally be spotted not only by the presence of the insects themselves but also by the products of their activities such as webbing, clumped-together food particles and holes in packaging materials.

Unless plastic films are laminated with foil or paper, insects are able to penetrate most of them quite easily, the rate of penetration usually being directly related to film thickness. In general, thicker films are more resistant than thinner films, and oriented films tend to be more effective than cast films. The looseness of the film has also been reported to be an important factor, loose films being more easily penetrated than tightly fitted films. Generally, the penetration varies depending on the basic resin from which the film is made, on the combination of materials, on the package structure, and of the species and stage of insects involved. The relative resistance to insect penetration of some flexible packaging materials is as follows:

— excellent resistance: polycarbonate; poly-ethylene-terephthalate;
— good resistance: cellulose acetate; polyamide; polyethylene (0.254 mm); polypropylene (biaxially oriented); poly-vinyl-chloride (unplasticised);
— fair resistance: acrylonitrile; poly-tetra-fluoro-ethylene; polyethylene (0.123 mm);
— poor resistance: regenerated cellulose; corrugated paper board; kraft paper; polyethylene (0.0254 - 0.100 mm); paper/foil/polyethylene laminate pouch; poly-vinylchloride (plasticised).

Some simple methods for obtaining insect resistance of packaging materials are as following:

— select a film and a film thickness that are inherently resistant to insect penetration;
— use shrink film over-wraps to provide an additional barrier;
— seal carton flaps completely.

Rodents

Rats and mice carry disease-producing organisms on their feet and/or in their intestinal tracts and are known to harbour salmonella of serotypes frequently associated with food-borne infections in humans. In addition to the public health consequences of rodent populations in close proximity to humans, these animals also compete intensively with humans for food. Rats and mice gnaw to reach sources of food and drink and to keep their teeth short. Their incisor teeth are so strong that rats have been known to gnaw through lead pipes and unhardened concrete, as well as sacks, wood and flexible packaging materials. Proper sanitation in food processing and storage areas is the most effective weapon in the fight against rodents, since all packaging materials apart from metal and glass containers can be attacked by rats and mice.

Raw Materials for Fruit and Vegetable Processing

Auxiliary raw materials used in fruit and vegetable processing technologies play a major role in the determination of their physical and chemical characteristics, sensory properties and nutritive value.

Water

Water is one of the essential factors in the activity of the processing centres; according to the final utilisation, water can be classified in three categories:

a) for technological utilisation (when it comes into direct contact with raw materials and enters in the finished product's composition),

b) for steam generators and

c) for receptacle cooling, washing of equipment and general hygiene.

Water coming into direct contact with the raw materials used for processing (washing, blanching, etc.) or that used as filling liquid of some canned products, must be of drinking water quality in terms of its physico-chemical and microbiological conditions. More important even than fulfilling drinking water standards, water used for these purposes pests certain specific characteristics related to the technological step or the raw material treated during the processing.

When very hard water is used for blanching vegetables some pecto-calcium and pectomagnesium complexes are formed which starts the hardening of vegetable tissues. This process continues over the pasteurisation of the finished product. When fruit is processed in sugar syrup, the use of hard water for the syrup preparation could induce the formation of a pectin-sugar-acid gel facilitated by the medium pH and presence of calcium salts.

Soft water has negative consequences associated with mineral and hydrosoluble substances and losses during blanching of vegetables. For some specific products such as peeled tomatoes, green beans and fine texture fruit, the addition of calcium salts (chlorure, sulphate, etc.) is employed to correct for low texture. The water hardness is an essential factor when used as filling liquid for canned products; ideally, the hardness of the water should be adapted to the raw material species used for canning. Thus a hardness of 3° is good for beans, 5° to 9° for green peas, green beans, and for fruit and vegetables with a tendency to disintegrate should use even harder water.

In the technological process of cucumber and gherkin preservation by natural acidification (lactic fermentation), water hardness has a paramount role. The literature maintains that as far as texture is concerned, the best results are obtained by using very hard water (about 30°); but since high magnesium and salt content has a negative effect on taste, in practice it is recommended that a water hardness of about 10° be used, which gives satisfactory results from both points of view.

Oxygen present in water can act as a corrosion factor in metal receptacles but this negative influence can be eliminated by preliminary boiling. An important factor is pH. Water for canning must be neutral or slightly basic. Acid water plays a major role in corrosion which is evident both on receptacles and on iron or copper equipment, where changes of product colour will be induced. More dangerous is the attack on lead pipes or to the mix used for can sticking and this can render the product toxic. For these reasons an acid water must be neutralised before use.

Water for steam generators

Two main conditions must be fulfilled:

— hardness has to be as low as possible, even zero, because precipitation of calcium salts can lead to the formation of encrustations ("crusts") in pipes and on equipment walls. For this reason, water treatment is practiced in the majority of installations;

— from a bacteriological point of view, the iron-bacteria must be eliminated with biological filters or oxidising substances. This is necessary in order to avoid iron hydroxide formed during equipment exploitation to deposit on the inner walls of the pipes.

The elimination of iron-bacteria is also of importance for the water used in processing steps.

Water for receptacle cooling and general hygiene

This should be of drinking water standard. Where this is not available in sufficient quantity, the use of industrial water is acceptable but only for cleaning of production rooms/workshops.

Sweeteners

Sugar

Sugar is the conventional name applied to sucrose. Physically there is icing, granulated and lump sugar. In fruit and vegetable processing, sugar is used only in its granulated form; this quality must be in the form of uniform crystals, white, shining and completely soluble in water. Concentration of various sugar solutions can be rapidly measured by refractometer reading or with areometers graduated in various ways: Brix, Baumé, etc.

Sugar solubility in water is dependent upon temperature; for example, in order to obtain a saturated solution, one must dissolve 2040 g in one litre of water at 20°C and 4870 g at 100°C. Taking into account this temperature related solubility, in practice the majority of sugar solutions are prepared by heating the water. Water should be as soft as possible because the calcium salts can precipitate on boiling.

Corn syrup

Corn syrup is obtained industrially by acid or enzymatic starch hydrolysis, using as starting raw materials maize (corn) or potatoes. In fruit processing, mainly in the production of marmalades, it is possible to use corn syrup. The average composition of this corn syrup is of about 32-40% dextrose (glucose), about 40% dextrins and 18-20% moisture. Sweetening power is 50% compared with sucrose. In a 10%-20% proportion with sucrose, addition of corn syrup has certain advantages:

a) it improves the shine and texture of marmalade;

b) it prevents "sugaring" defect and

c) it reduces the too sweet taste of finished products obtained with sugar alone.

Salt

Salt is used in order to give to the finished products a specifically salty taste and as a preserving substance. From a chemical point of view the term salt means sodium chloride but in practice the product is never in a pure state.

The presence of a significant quantity of magnesium chloride increases the hygroscopicity, gives a bitter taste and can induce corrosion of receptacles. From a microbiological point of view, salt it is not a sterile product but on the contrary contains various micro-organisms, mainly halophil bacteria. Salt solubility is only slightly influenced by temperature.

Post-harvest Technology of Fresh Fruits

Once fruit is harvested, any natural resistance to the action of spoiling micro-organisms is lost. Changes in enzymatic systems of the fruit also occur on harvest which may also accelerate the activity of spoilage organisms. Means that are commonly used to prevent spoilage of fruits must include:

— care to prevent cutting or bruising of the fruit during picking or handling;

— refrigeration to minimise growth of micro-organisms and reduce enzyme activity;

— packaging or storage to control respiration rate and ripening;

— use of preservatives to kill micro-organisms on the fruit.

A principal economic loss occurring during transportation and/or storage of produce such as fresh fruit is the degradation which occurs between the field and the ultimate destination due to the effect of respiration. Methods to reduce such degradation are as follows:

— refrigerate the produce to reduce the rate of respiration;

— vacuum cooling;

— reduce the oxygen content of the environment in which the produce is kept to a value not above 5% of the atmosphere but above the value at which anaerobic respiration would begin.

Some recent developments in post-harvest technology of fresh fruits and vegetables are:

Harvest maturity: This is particularly important with fruit for export. One recent innovation is the measurement of resonant frequency of the fruit which should enable the grading out of over mature and under-mature fruit before they are packed for export.

Harvest method: Considerable research is continuing on mechanical harvesting of perishable crops with a view to minimising damage. In fruit trees, controlling their height by use of dwarfing rootstocks, pruning and growth regulating chemicals will lead to easier, cheaper more accurate harvesting.

Handling systems: Field packing of various vegetables for export has been carried out for many years. In the last decade or so this has been applied, in selected cases, to a few tropical fruit types. Where this system can be practiced it has considerable economic advantages in saving the cost of building, labour and equipment and can result in lower levels of damage into crops.

Pre-cooling: Little innovation has occurred in crop pre-cooling over the last decade. However high velocity, high humidity forced air systems have continued to be developed and refined. These are suitable for all types of produce and are relatively simple to build and operate and, while not providing the speed of cooling of a vacuum or hydrocooler, have the flexibility to be used with almost all crops.

Chemicals: There is a very strong health lobby whose objective is to reduce the use of chemicals in agriculture and particularly during the post harvest period. Every year sees the prohibition of the use of commonly used post-harvest chemicals. New ways need to be developed to control post-harvest diseases, pest and sprouting.

Coatings: Slowing down the metabolism of fruit and vegetables by coating them with a material which affects their gaseous exchange is being tested and used commercially on a number of products.

Controlled environment transport: Recent innovations in this technique have produced great progress as a result of the development and miniaturisation of equipment to measure carbon dioxide and oxygen. Several companies now offer containers where the levels of these two gases can be controlled very precisely.

Preservation Methods for Fruit and Vegetables

Drying/Dehydration and Concentration

Water and water activity

Micro-organisms in a healthy growing state may contain in excess of 80% water. They get this water from the food in which they grow. If the water is removed from the food it also will transfer out of the bacterial cell and multiplication will stop. Partial drying will be less effective than total drying, though for some micro-organisms partial drying may be quite sufficient to arrest bacterial growth and multiplication. Bacteria and yeasts generally require more moisture than moulds, and so moulds often will be found growing on semi-dry foods where bacteria and yeasts find conditions unfavourable; example are moulds growing on partially dried fruits.

Slight differences in relative humidity in the environment in which the food is kept or in the food package can make great differences in the rate of micro-organism multiplication. Since micro-organisms can live in one part of a food that may differ in moisture and other physical and chemical conditions from the food just millimetres away, we must be concerned with conditions in the "microenvironment" of the micro-organisms. Thus it is common to refer to water conditions in terms of specific activity.

The term "water activity" is related to relative humidity. Relative humidity is defined as the ratio of the partial pressure of water vapour in the air to the vapour pressure of pure water at the same temperature. Relative humidity refers to the atmosphere surrounding a material or solution. Water activity or aw is a property of solutions and is the ratio of vapour pressure of the solution compared with the vapour pressure of pure water at the same temperature. Under equilibrium conditions water activity equals:

$$a_w = RH / 100$$

When we speak of moisture requirements of micro-organisms we really mean water activity in their immediate environment, whether this be in solution, in a particle of food or at a surface in contact with the atmosphere. At the usual temperatures permitting microbial growth, most bacteria require a water activity in the range of about 0.90 to 1.00. Some yeasts and moulds grow slowly at a water activity down to as low as about 0.65.

Qualitatively, water activity is a measure of unbound, free water in a system available to support biological and chemical reactions. Water activity, not absolute water content, is what bacteria, enzymes and chemical reactants encounter and are affected by at the micro-environmental level in food materials.

Two foods with the same water content can have very different a_w values depending upon the degree to which water is free or otherwise bound to food constituents. When a food is in moisture equilibrium with its environment, then the a_w of the food will be quantitatively equal to the RH divided by 100. Qualitatively, water activity is a measure of free or available water, to be distinguished from unavailable or bound water. These states of water also bear a relationship to the characteristic sigmoid shapes of water absorption isotherm curves of various foods. Thus, according to theory, most of the water corresponding to the portion of the curve below its first inflection point is believed to be tightly bound water, often referred to as an adsorbed mono-molecular layer of water.

Moisture corresponding to the region above this point and up to the curve's second inflection point is thought to exist largely as multi-molecular layers of water less tightly held to food constituent surfaces. Beyond this second inflection point moisture generally is considered to be largely free water condensed in capillaries and interstices within the food. In this latter portion of the sorption isotherm curve small changes in moisture content result in great changes in a food's a_w.

The technique of drying is probably the oldest method of food preservation practiced by mankind. The removal of moisture prevents the growth and reproduction of micro-organisms causing decay and minimises many of the moisture mediated deterioration reactions. It brings about substantial reduction in weight and volume minimising packing, storage and transportation costs and enable storability of the product under ambient temperatures, features especially important for developing countries. The sharp rise in energy costs has promoted a dramatic upsurge in interest in drying world-wide over the last decade.

Dehydration involves the application of heat to vaporise water and some means of removing water vapour after its separation from the fruit/ vegetable tissues. Hence it is a combined/simultaneous transfer operation for which energy must be supplied. A current of air is the most common medium for transferring heat to a drying tissue and convection is mainly involved. The two important aspects of mass transfer are:

— the transfer of water to the surface of material being dried and

— the removal of water vapour from the surface.

In order to assure products of high quality at a reasonable cost, dehydration must occur fairly rapidly. Four main factors affect the rate and total drying time:

— the properties of the products, especially particle size and geometry;

— the geometrical arrangement of the products in relation to heat transfer medium (drying air);

— the physical properties of drying medium/ environment;

— the characteristics of the drying equipment.

It is generally observed with many products that the initial rate of drying is constant and then decreases, sometimes at two different rates. The drying curve is divided into the constant rate period and the falling rate period.

Surface area: Generally the fruit and vegetables to be dehydrated are cut into small pieces or thin layers to speed heat and mass transfer. Subdivision speeds drying for two reasons:

— large surface areas provide more surface in contact with the heating medium (air) and more surface from which moisture can escape;

— smaller particles or thinner layers reduce the distance heat must travel to the centre of the food and reduce the distance through which moisture in the centre of the food must travel to reach the surface and escape.

Temperature: The greater the temperature difference between the heating medium and the food the greater will be the rate of heat transfer into the food, which provides the driving force for moisture removal. When the heating medium is air, temperature plays a second important role. As water is driven from the food in the form of water vapour it must be carried away, or else the moisture will create a saturated atmosphere at the food's surface which will slow down the rate of subsequent water removal. The hotter the air the more moisture it will hold before becoming saturated. Thus, high temperature air in the vicinity of the dehydrating food will take up the moisture being driven from the food to a greater extent than will cooler air. Obviously, a greater volume of air also can take up more moisture than a lesser volume of air.

Air velocity: Not only will heated air take up more moisture than cool air, but air in motion will be still more effective. Air in motion, that is, high velocity air, in addition to taking up moisture will sweep it away from the drying food's surface, preventing the moisture from creating a saturated atmosphere which would slow down subsequent moisture removal. This is why clothes dry more rapidly on a windy day. Some other phenomena influence the drying process and a few elements are summarised below.

Dryness of air: When air is the drying medium of food, the drier the air the more rapid is the rate of drying. Dry air is capable of absorbing and holding moisture. Moist air is closer to saturation and so can absorb and hold less additional moisture than if it were dry. But the dryness of the air also determines how low a moisture content the food product can be dried to.

Atmospheric pressure and vacuum: If food is placed in a heated vacuum chamber the moisture can be removed from the food at a lower temperature than without a vacuum. Alternatively, for a given temperature, with or without vacuum, the rate of water removal from the food will be greater in the vacuum. Lower drying temperatures and shorter drying times are especially important in the case of heat-sensitive foods.

Evaporation and temperature: As water evaporates from a surface it cools the surface. The cooling is largely the result of absorption by the water of the latent heat of phase change from liquid to gas. In doing this the heat is taken from the drying air or the heating surface and from the hot food, and so the food piece or droplet is cooled.

Time and temperature: Since all important methods of food dehydration employ heat, and food constituents are sensitive to heat, compromises must be made between maximum possible drying rate and maintenance of food quality. As is the case in the use of heat for pasteurisation and sterilisation, with few exceptions drying processes which employ high temperatures for short times do less damage to food than drying processes employing lower temperatures for longer times. Thus, vegetable pieces dried in a properly designed oven in four hours would retain greater quality than the same products sun dried over two days. Several drying processes will achieve dehydration in a matter of minutes or even less if the food is sufficiently subdivided.

Several types of dryers and drying methods, each better suited for a particular situation, are commercially used to remove moisture from a wide variety of food products including fruit and vegetables. While sun drying of fruit crops is still practiced for certain fruit such as prunes, figs, apricots, grapes and dates, atmospheric dehydration processes are used for apples, prunes, and several vegetables; continuous processes as tunnel, belt trough, fluidised bed and foam-mat drying are mainly used for vegetables.

Spray drying is suitable for fruit juice concentrates and vacuum dehydration processes are useful for low moisture / high sugar fruits like peaches, pears and apricots. Factors on which the selection of a particular dryer/ drying method depends include:

— form of raw material and its properties;

— desired physical form and characteristics of dried product;

— necessary operating conditions;

— operation costs.

There are three basic types of drying process:

— sun drying and solar drying;

— atmospheric drying including batch (kiln, tower and cabinet dryers) and continuous (tunnel, belt, belt-trough, fluidised bed, explosion puff, foam-mat, spray, drum and microwave);

— sub-atmospheric dehydration (vacuum shelf/belt/drum and freeze dryers).

The scope has been expanded to include use of low temperature, low energy process like osmotic dehydration. As far dryers are concerned, one useful division of dryer types separates them into air convection dryers, drum or roller dryers, and vacuum dryers. Using this breakdown, Table 3 indicates the applicability of the more common dryer types to liquid and solid type foods.

Table 3: Common dryer types used for liquid and solid foods.

Dryer type	*Usual food type*
Air convection dryers	
kiln	pieces
cabinet, tray or pan	pieces, purées, liquids
tunnel	pieces
continuous conveyor belt	purées, liquids
belt trough	pieces
air lift	small pieces, granules
fluidised bed	small pieces, granules
spray	liquid, purées
Drum or roller dryers	
atmospheric	purées, liquids
vacuum	purées, liquids
Vacuum dryers	
vacuum shelf	pieces, purées, liquids
vacuum belt	purées, liquids
freeze dryers	pieces, liquids

Natural drying

Surplus production and specifically grown crops may be preserved by natural drying for use until the next crop can be grown and harvested. Natural dried products can also be transported cheaply for distribution to areas where there are permanent shortages of fruit and vegetables. The methods of producing sun and solar dried fruit and vegetables described here are simple to carry out and inexpensive. They can be easily employed by grower, farmer, cooperative, etc.

The best time to preserve fruits and vegetables is when there is a surplus of the product and when it is difficult to transport fresh materials to other markets. This is especially true for crops which are very easily damaged in transport and which stay in good condition for a very short time. Preservation extends the storage (shelf) life of perishable foods so that they can be available throughout the year despite their short harvesting season.

Sun and solar drying of fruits and vegetables is a cheap method of preservation because it uses the natural resource/ source of heat: sunlight. This method can be used on a commercial scale as well at the village level provided that the climate is hot, relatively dry and free of rainfall during and immediately after the normal harvesting period. The fresh crop should be of good quality and as ripe (mature) as it would need to be if it was going to be used fresh. Poor quality produce cannot be used for natural drying.

Dried fruit and vegetables have certain advantages over those preserved by other methods. They are lighter in weight than their corresponding fresh produce and, at the same time, they do not require refrigerated storage. However, if they are kept at high temperatures and have a high moisture content they will turn brown after relatively short periods of storage. Different lots at various stages of maturity (ripeness) must not be mixed together; this would result in a poor dried product. Some varieties of fruit and vegetables are better for natural drying than other; they must be able to withstand natural drying without their texture becoming tough so that they are not difficult to reconstitute.

Some varieties are unsuitable because they have irregular shape and there is a lot of wastage in trimming and cutting such varieties. Damaged parts which have been attacked by insects, rodents, diseases, etc. and parts which have been discoloured or have a bad appearance or colour, must be removed. Before trimming and cutting, most fruit and vegetables must be washed in clean water. Onions are washed after they have been peeled.

Trimming includes the selection of the parts which are to be dried, cutting off and disposing of all unwanted material. After trimming, the greater part of the fruit and vegetables cut into even slices of about 3 to 7 mm thick or in halves/quarters, etc. It is very important to have all slices/ parts in one drying lot of the same thickness/size; the actual thickness will depend on the kind of material. Uneven slices or different sizes dry at different rates and this result in a poor quality end product. Onions and root crops are sliced with a hand slicer or vegetable cutter; bananas, tomatoes and other vegetables or fruit are sliced with stainless-steel knives.

As a general rule plums, grapes, figs, dates are dried as whole fruits without cutting/slicing. Some fruit and vegetables, in particular bananas, apples and potatoes, go brown very quickly when left in the air after peeling or slicing; this discoloration is due to an active enzyme called phenoloxidase. To prevent the slices from going brown they must be kept under water until drying can be started. Salt or sulphites in solution give better protection. However, whichever method is used, further processing should follow as soon as possible after cutting or slicing. Blanching—exposing fruit and vegetable to hot or boiling water—as a pre-treatment before drying has the following advantages:

— it helps clean the material and reduce the amount of micro-organisms present on the surface;

— it preserves the natural colour in the dried products; for example, the carotenoid (orange and yellow) pigments dissolve in small intracellular oil drops during blanching and in this way they are protected from oxidative breakdown during drying;

— it shortens the soaking and/or cooking time during reconstitution.

During hot water blanching, some soluble constituents are leached out: water-soluble flavours, vitamins (vitamin C) and sugars. With potatoes this may be an advantage as leaching out of sugars makes the potatoes less prone to turning brown. Blanching is a delicate processing step; time, temperature and the other conditions must be carefully monitored. A suitable water-blanching method in traditional processing is as follows:

— the sliced material is placed on a square piece of clean cloth; the corners of the cloth are tied together;

— a stick is put through the tied corners of the cloth;

— the cloth is dipped into a pan containing boiling water and the stick rests across the top of the pan thus providing support for the cloth bag.

The average blanching time is 6 minutes. The start of blanching has to be timed from the moment the water starts to boil again after the cloth bag has been dipped into the pan. While the material is being blanched the cloth bag should be raised and lowered in the water so that the material is heated evenly. When the blanching time is completed the cloth bag and its content should be dipped into cold water to prevent over-blanching. If products are over-blanched (boiled for too long) they will stick together on the drying trays and they are likely to have a poor flavour.

Green beans, carrots, okra, turnip and cabbage should always be blanched. The producer can choose whether or not potatoes need blanching. Blanching is not needed for onions, leeks, tomatoes and sweet peppers. Tomatoes are dipped into hot water for one minute when they need to be peeled but this is not blanching. As a rule fruit is not blanched.

Use of preservatives

Preservatives are used to improve the colour and keeping qualities of the final product for some fruits and vegetables. Preservatives include items such as sulphur dioxide, ascorbic acid, citric acid, salt and sugar and can either be simple or compound solutions. Treatment with preservatives takes place after blanching or, when blanching is not needed, after slicing. In traditional, simple processing the method recommended is:

— put enough preservative solution to cover the cloth bag into a container/ pan;

— dip the bag containing the product into the preservative solution for the amount of time specified;

— remove the bag and put it on a clean tray while the liquid drains out. The liquid which drains out must not go back into the preservative solution because it would weaken the solution.

Care must be taken after each dip to refill the container to the original level with fresh preservative solution of correct strength. After five lots of material have been dipped, the remaining solution is thrown away; *ie*. a fresh lot of preservative solution is needed for every 5 lots of material. The composition and strength of the preservative solution vary for different fruit and vegetables. The strength of sulphur dioxide is expressed as "parts per million" (ppm). 1.5 grams of sodium metabisulphite in one litre of water gives 1000 ppm of sulphur dioxide. Details for solutions of different strengths are given in the following table.

Table 4: Dilutions of sodium metabisulphite with water to obtain "PM" of sulphur dioxide (SO_2)

PPM SO_2	*Sodium Metabisulphite*	
	Grams per litre of water	*Grams per 20 litre tin of water*
1000	1.5	30.0
2000	3.0	60.0
3000	4.5	90.0
4000	6.0	120.0
5000	7.5	150.0
6000	9.0	180.0
7000	10.5	210.0

One level teaspoon of sodium metabisulphite = c. 5 g.

Sodium bicarbonate is added to the blanching water when okra, green peas and some other green vegetables are blanched. The chemical raises the pH of the blanching water and prevents the fresh green colour of chlorophyll being changed into pheophytin which is unattractive brownish-green. The preservative solutions in the fruit and vegetable pre-treatment can only be used in enamelled, plastic or stainless-steel containers; never use ordinary metal because solutions will corrode this type of container. As a general rule, preservatives are not used for treating onions, garlic, leeks, chilies and herbs.

Osmotic dehydration

In osmotic dehydration the prepared fresh material is soaked in a heavy (thick liquid sugar solution) and/or a strong salt solution and then the material is sun or solar dried. During osmotic treatment the material loses some of its moisture. The syrup or salt solution has a protective effect on colour, flavour and texture. This protective effect remains throughout the drying process and makes it possible to produce dried products of high quality. This process makes little use of sulphur dioxide.

Sun drying

The main problems for sun drying are dust, rain and cloudy weather.

Therefore, drying areas should be dust-free and whenever there is a threat of a dust storm or rain, the drying trays should be stacked together and placed under cover. In order to produce dust-free and hygienically clean products, fruit and vegetable material should be dried well above ground level so that they are not contaminated by dust, insects, livestock or people. All materials should be dried on trays designed for the purpose; the most common drying trays have wooden frames with a fitted base of nylon mosquito netting.

Mesh made of woven grass can also be used. Metal netting must NOT be used because it discolours the product. The trays should be placed on a framework at table height from the ground. This allows the air to circulate freely around the drying material and it also keeps the food product well away from dirt. Ideally the area should be exposed to wind and this speed up drying, but this can only be done if the wind is free of dust. With 80 cm x 50 cm trays, the approximate load for a tray is 3 kg; the material should be spread in even layers.

During the first part of the drying period, the material should be stirred and turned over at least once an hour. This will help the material dry faster and more evenly, prevent it sticking together and improve the quality of the finished product. Products for sun drying should be prepared early in the day; this will ensure that the material enjoys the full effect of the sun during the early stages of drying. At night the trays should be stacked in a ventilated room or covered with canvas. Plastic sheets should never be used for covering individual trays during sun drying. Dry or nearly dry products can be blown out of the tray by the wind. However, this can be protected by covering the loaded tray with an empty one; this also gives protection against insects and birds.

Shade drying

Shade drying is carried out for products which can lose their colour and/or turn brown if put in direct sunlight. Products which have naturally vivid colours like herbs, green and red sweet peppers, chilies, green beans and okra give a more attractive end-product when they are dried in the shade. The principles for the shade drying are the same as for sun drying. The material to be dried requires full air circulation. Therefore, shade drying is carried out under a roof or thatch which has open sides; it cannot be done either inside conventional buildings with side walls or in compounds sheltered from wind. Under dry conditions when there is a good circulation of air, shade drying takes little more time than is normally required for drying in full sunlight.

In the selection of appropriate solar dryers for commercial scale operation, it is imperative that economics be kept in view at all time. A total Energy System concept should be employed and due consideration be given to parasitic energy consumption. The following features have been identified:

— large scale dryers are more promising than small scale ones. However, small scale dryers should not be neglected.

— the dryer should be designed to maximise the utilisation factor of the capital investment, *ie*. multi-products (fruit, vegetables and other raw material) and multi-use (*eg*. drying and heating water for domestic use).

— in general, an auxiliary heat source should be provided to assure reliability, to handle peak loads and also to provide continuous drying during periods of no sunshine.

— forced convection indirect dryers are preferred because they offer better control, more uniform drying and because of their high heat collection efficiency result in smaller collector area. However, parasitic power should be kept to a minimum.

Two dryer systems have been identified:

— a cabinet type dryer with natural convection for internal air circulation for the processing of dried fruit such as mango, banana, pineapple, apricot, pear, apple, etc. and also for potato chips and other vegetables;

— a greenhouse type dryer with forced air circulation.

Some of the barriers to the commercial development of solar dryers have been attributed to:

— initial cost—poor farmers cannot afford them

— durability—constant breakdown due to using low cost building materials

— misuse—through lack of training and technical skills

— dependability and reliability—during the wet season when drying is critical there is not enough solar energy available

— the wider use of Solar Drying Systems has been limited by other factors which are not necessarily of a technical or technological nature. Among the most important are the lack of national policies directed to promoting the drying of produce at the production site, in order to reduce losses, improve quality and increase farmers' earnings.

In the case of simple natural convection dryers it may be more appropriate to build and operate a number of small units. Multiplicity allows diversity, since more than one crop can be dried at a time. A further advantage is that if one dryer is out of operation due to damage, drying can still continue at reduced capacity using the other dryers. On the other hand, more sophisticated dryers, such as forced convection solar dryers, benefit from economies of scale due to the investment tied up in the fan and the source of heat.

Generally speaking, one large dryer will be more cost-effective than two smaller units. However, it should be taken into consideration that an oversized unit will be operating at less than full capacity, reducing any cost advantage. The drying area required will depend on local conditions, commodity and number of trays on each rack or trolley.

Construction methods and available materials may vary considerably from location to location. Some general guidelines regarding factors which must be considered can, however, be given:

Dimensions of standard materials: Where possible, design should take account of the sizes of material locally available. For example, it would be poor design to specify the width of a corrugated iron collector as 1.1 m if the standard width of a corrugated iron sheet is 1 m. Before finalising a design the commercial availability of materials must be ascertained.

Use of rural materials: The cost of building of solar dryer can be minimised if the producer is able to use wood cut straight from the forest rather than prepared timber. Careful design in the development stage of a dryer can often facilitate the use of cheaper materials. Difficulties caused by these materials are in joining pieces of the structure, in sealing the structure against air leaks, and in attaching the plastic sheet to the (wooden) frame. There is obvious scope for designs which use prepared timber for strategic points and unprepared at others. Where the use of wood is necessary, remember to take environmental factors into consideration. For example, determine the effect of flash flooding or termites might be and take the appropriate preventive action.

Use of plastic sheets: For many solar dryers, the clear plastic sheet used is the major capital cost to the farmer; therefore, the type of plastic chosen is important. A choice must be made between a relatively cheap plastic such as ordinary polyethylene which will last, at best, for one season due to photo-degradation and wear and tear; and a more expensive, better quality plastic less prone to photo-degradation; or even glass or a rigid plastic.

Attaching plastic sheet to the framework structure, so as to minimise the likelihood of the plastic being torn is, perhaps, the most difficult part of building a dryer. Listed below are some general points which should be followed to prolong the useful lifetime of plastic sheet on a solar dryer:

— when attaching plastic sheet to the framework, care should be taken to stretch the plastic at the points of attachment, but the plastic should not be so loose that it will flap about in the wind;

— rather than merely stapling or nailing the plastic directly to the framework, it is preferable to sandwich the plastic between the framework and a batten. This may not be practical when unprepared wood or other materials are being used.

— no sharp edges should come in contact with the plastic sheet since these will initiate tears;

— fold over the plastic at the point of attachment to the frame, so that there are two or more layers of plastic. This will help prevent tears;

— when fixing the sheet over the framework, sags and hollows in which water can collect should be avoided wherever possible;

— the dryer should be handled as carefully and as seldom as possible during operation and when not in use.

The following design factors must be established:

— the throughput of the dryer over the productive season; the size of batch to be dried;

— the drying period(s) under stated conditions;

— the initial and desired final moisture content of the commodity (if known);

— the drying characteristics of the commodity, such as maximum drying temperature, effect of sunlight upon the product quality, etc.;

— climatic conditions during the drying season, *ie*. sunlight intensity and duration; air temperature and humidity; wind speed (such data may be available from local meteorological stations);

— availability and reliability of electrical power;
— the availability, quality, durability and price of potential construction materials such as:
  — glazing materials: glass, plastic sheet or film;
  — wood (prepared or unprepared);
  — nails, screw, bolts, etc.;
  — metal sheet, flat or corrugated angle iron;
  — bricks (burnt or mud), concrete blocks, stones, cement, sand, etc.
  — roofing thatch;
  — metal mesh, wire netting, etc.
  — mosquito netting, muslin, etc.
  — bamboo or fibre weave;
  — black paint, other blackening materials;
  — insulation material; sawdust, etc.;
— the type of labour available to build and operate the dryer;
— the availability of clean water at the site for preparation of the commodity prior to drying.

In any one situation there may well be other technical factors that need to be considered.

From the initial considerations, estimates of the capital costs of the dryer, the price of the commodity to be dried, and the likely selling price of the dried product will have been made. Other question that need to be considered are the following:

— who will own the dryer?
— is the dryer to be constructed by the end-user (with or without advice from extension agencies), local contractors, or other organisations?
— who will operate and maintain it?
— how can the drying operation be incorporated into current practices?
— are sources of finance from local authorities or extension agencies available, etc.?

Obviously there are many other socio-economic factors, particularly those of a local nature, which must be taken into account. It cannot be stressed too highly that if such factors are not taken into account and evaluated, then is every chance that an inappropriate dryer design may result. Equal emphasis must be placed on both technical and socio-economic factors.

Drying tray

The drying tray described requires seasoned timber 22.5 mm thick x 50 mm wide. A sun drying tray requires 6 meters of seasoned timber 22.5 mm thick x 50 mm wide. Cut the timber into lengths of 900 mm long for the sides of the tray and 600 mm long for the ends—4 pieces of each length will be needed. The ends of each piece are cut as shown in the drawing—this is to make flush fitting joints. Join the corners using small brass screws 20 mm long. To make extra strong joints use good quality wood glue as well as the screws.

The nylon mosquito netting or grass woven mesh can be fitted between the frames as shown in the bottom drawing. Cut the mesh a little larger than the size of the frame. Using drawing pins, pin the mesh to the outside edges of one of the frames—the mesh should be pulled tight as the pins are put in around the edges. Lay the other frame on top and drill holes about 3 mm in diameter at the points marked X in the top drawing. Use nails that are a tight fit in the holes and tap gently into place leaving a portion standing above the frame. Cut off the standing part to leave a piece about 12 mm long which is then bent over and tapped firmly down onto the frame. When the frame has been put together tightly, the drawing pins can be removed.

Preservation by Concentration

Foods are concentrated for many of the same reasons that they are dehydrated; concentration can be a form of preservation but this is true only for some foods. Concentration reduces weight and volume and results in immediate economic advantages. Nearly all liquid foods which are dehydrated are concentrated before they are dried. This is because in the early stages of water removal, moisture can be more economically removed in highly efficient evaporators than in dehydration equipment. Further, increased viscosity from concentration often is needed to prevent liquids from running off drying surfaces or to facilitate foaming or puffing.

Foods are also concentrated because the concentrated forms have become desirable components of diet in their own right. Thus, fruit juices plus sugar with concentration becomes jelly. The more common concentrated fruit and vegetable products include items as fruit and vegetable juices and nectars, jams and jellies, tomato paste, many types of fruit purées used by bakers, candy makers and other food manufacturers. The level of water in virtually all concentrated foods is in itself more than enough to permit microbial growth. Yet while many concentrated foods such as non-acid fruit and vegetable purées may quickly undergo microbial spoilage unless additionally processed, such items as sugar syrups, jellies and jams are relatively "immune" to spoilage; the difference of course is in what is dissolved in the remaining water and what osmotic concentration is reached.

Table 5: Specific gravity and solids. Tomato pulp and commercial tomato concentrates

Tomato Solids,%	*Specific gravity at 68°F (20°C)*	*Dry Tomato Solids*	
		per Gal. at 68°F, lb	*per Litre at 20°C, g*
6.0 Tomato pulp	1.025	0.51	61
10.8	1.045	0.94	113
12.0 Tomato purée	1.050	1.05	126
14.2	1.060	1.25	151
16.5	1.070	1.47	177
25.0	1.107	2.31	277
26.0	1.112	2.41	289
28.0 Tomato paste	1.120	2.61	314
30.0	1.129	2.82	339
32.0	1.138	3.03	364

Sugar and salt in concentrated solutions have high osmotic pressure. When these are sufficient to draw water from microbial cells or prevent normal diffusion of water into these cells, a preservative condition exists. The critical concentration of sugar in water to prevent microbial growth will vary depending upon the type of micro-organisms and the presence of other food constituents, but usually 70% sucrose in solution will stop growth of all micro-organisms in foods. Less than this concentration may be effective but for short periods of time unless the foods contain acid or they are refrigerated.

Salt becomes a preservative when its concentration is increased and levels of about 18% to 25% in solution generally will prevent all growth of micro-organisms in foods. Except in the case of certain briny condiments, however, this level is rarely tolerated in foods. Removal of water by concentration also increases the level of food acids in solution (particularly significant in concentrated fruit juices).

While the preservation effects of food concentration are important, the main reason of most food concentration is to reduce food weight and bulk. Tomato pulp which is ground tomato minus the skins and seeds, has a solid content of only 6 % and so a 3.785 litre can would contain only 231 g of tomato solids.

Concentrated to 32% solids, the same can would contain 1.38 kg of tomato solids or six times the value of product. For a manufacturer needing tomato solids such a producer of soups, canned spaghetti or frozen pizza the saving from concentration are enormous in cans, transportation costs, warehousing costs and handling costs throughout his operation.

Concentration methods

a) *Solar concentration*: As in food dehydration, one of the simplest methods of evaporating water is with solar energy. A typical example of this method is production at farm level in developing countries of fruit pastes/leathers (such as apricot or plum pastes).

b) *Open Kettles*: Some foods can be satisfactorily concentrated in open kettles that are heated by steam. This is the case for jellies and jams, tomato juices and purées and for certain types of soups. High temperatures and long concentration times should be avoided in order to reduce or eliminate damage. It is also necessary to avoid thickening and burn-on of product to the kettle wall as these gradually lower the efficiency of heat transfer and slow the concentration process.

However, when the process is under control, this type of evaporation is still highly recommended for small scale operations in developing countries. It is a quite widely used system, mainly for jellies, jams and marmalades.

c) Flash evaporators.

d) Thin-film Evaporators.

e) Vacuum evaporators. It is common to construct several vacuumised vessels in series so that the product moves from one vacuum chamber to the next and thereby becomes progressively more concentrated in stages. With such an arrangement the successive stages are maintained at progressively higher degrees of vacuum, and the hot water vapour produced by the first stage is used to heat the second stage, the vapour from the second stage heats the third stage and so on. In this way maximum use of heat energy is made. Such system is called a multiple effect vacuum evaporator.

f) Freeze Concentration. This process has been known for many years and has been applied commercially to orange juice. However, high processing costs due largely to losses of juice occlude [unclear] to the ice crystals, have limited the number of installations to date.

g) Ultrafiltration and reverse Osmosis.

Obviously concentration that exposes food to 100°C or higher temperatures for prolonged periods can cause major changes in organoleptic and nutritional properties. Cooked foods and darkening of colour are two of the more common heat induced results which must be kept under control during a well designed process with an efficient evaporator which is still "safe". Microbial destruction is another type of change that may occur during concentration and will be largely dependent upon temperature.

Concentration at a temperature of 100°C or slightly above will kill many micro-organisms but cannot be depended on to destroy bacterial spores. When the food contains acid, such as fruit juices, the kill will be greater but again sterility is unlikely. On the other hand, when concentration is done under vacuum many bacterial types not only survive the low temperatures but multiply in the concentrating equipment. It is therefore necessary to stop frequently and sanitise low temperature evaporators and where sterile concentrated foods are required, to resort to an additional preservation treatment.

Chemical Preservation

Many chemicals will kill micro-organisms or stop their growth but most of these are not permitted in foods. Chemical food preservatives are those substances which are added in very low quantities (up to 0.2%) and which do not alter the organoleptic and physico-chemical properties of the foods at or only very little.

Preservation of food products containing chemical food preservatives is usually based on the combined or synergistic activity of several additives, intrinsic product parameters (*eg*. composition, acidity, water activity) and extrinsic factors (*eg*. processing temperature, storage atmosphere and temperature). This approach minimises undesirable changes in product properties and reduces concentration of additives and extent of processing treatments. The concept of combinations of preservatives and treatments to preserve foods is frequently called the hurdle or barrier concept.

Combinations of additives and preservatives systems provide unlimited preservation alternatives for applications in food products to meet consumer demands for healthy and safe foods. Chemical food preservatives are applied to foods as direct additives during processing, or develop by themselves during processes such as fermentation. Certain preservatives have been used either accidentally or intentionally for centuries, and include sodium chloride (common salt), sugar, acids, alcohols and components of smoke.

In addition to preservation, these compounds contribute to the quality and identity of the products, and are applied through processing procedures such as salting, curing, fermentation and smoking. Traditional chemical food preservatives and their use in fruit and vegetable processing technologies are:

— common salt: brined vegetables;
— sugars (sucrose, glucose, fructose and syrups):
— foods preserved by high sugar concentrations: jellies, preserves, syrups, juice concentrates;
— interaction of sugar with other ingredients or processes such as drying and heating;
— indirect food preservation by sugar in products where fermentation is important (naturally acidified pickles and sauerkraut).

Acidulants and other preservatives formed in or added to fruit and vegetable products are as follows:

Lactic acid: This acid is the main product of many food fermentations; it is formed by microbial degradation of sugars in products such as sauerkraut and pickles. The acid produced in such fermentations decreases the pH to levels unfavourable for growth of spoilage organisms such as putrefactive anaerobes and butyric-acid-producing bacteria. Yeasts and moulds that can grow at such pH levels can be controlled by the inclusion of other preservatives such as sorbate and benzoate.

Acetic acid: Acetic acid is a general preservative inhibiting many species of bacteria, yeasts and to a lesser extent moulds. It is also a product of the lactic-acid fermentation, and its preservative action even at identical pH levels is greater than that of lactic acid. The main applications of vinegar (acetic acid) includes products such as pickles, sauces and ketchup.

Acidulants

— Malic and tartaric (tartric) acids is used in some countries mainly to acidify and preserve fruit sugar preserves, jams, jellies, etc.

— Citric acid is the main acid found naturally in citrus fruits; it is widely used (in carbonated beverages) and as an acidifying agent of foods because of its unique flavour properties. It has an unlimited acceptable daily intake and is highly soluble in water. It is a less effective antimicrobial agent than other acids.

— Ascorbic acid or vitamin C, its isomer isoascorbic or erythorbic acid and their salts are highly soluble in water and safe to use in foods.

Lipophilic acid food preservatives

Benzoic acid in the form of its sodium salt, constitutes one of the most common chemical food preservative. Sodium benzoate is a common preservative in acid or acidified foods such as fruit juices, syrups, jams and jellies, sauerkraut, pickles, preserves, fruit cocktails, etc. Yeasts are inhibited by benzoate to a greater extent than are moulds and bacteria.

Sorbic acid is generally considered non toxic and is metabolised; among other common food preservatives the WHO has set the highest acceptable daily intake (25 mg/kg body weight) for sorbic acid. Sorbic acid and its salts are practically tasteless and odourless in foods, when used at reasonable levels (< 0.3 %) and their antimicrobial activity is generally adequate.

Sorbates are used for mould and yeast inhibition in a variety of foods including fruits and vegetables, fruit juices, pickles, sauerkraut, syrups, jellies, jams, preserves, high moisture dehydrated fruits, etc. Potassium sorbate, a white, fluffy powder, is very soluble in water (over 50%) and when added to acid foods it is hydrolysed to the acid form. Sodium and calcium sorbates also have preservative activities but their application is limited compared to that for the potassium salt, which is employed because of its stability, general ease of preparation and water solubility.

Gaseous preservatives

Sulphur dioxide and sulphites. Sulphur dioxide (SO_2) has been used for many centuries as a fumigant and especially as a wine preservative. It is a colourless, suffocating, pungent-smelling, non-flammable gas and is very soluble in cold water (85 g in 100 ml at 25°C). Sulphur dioxide and its various sulphites dissolve in water, and at low pH levels yield sulphurous acid, bisulphite and sulphite ions. The various sulphite salts contain 50-68% active sulphur dioxide. A pH dependent equilibrium is formed in water and the proportion of SO_2 ions increases with decreasing pH values. At pH values less than 4.0 the antimicrobial activity reaches its maximum.

Sulphur dioxide is used as a gas or in the form of its sulphite, bisulphite and metabisulphite salts which are powders. The gaseous form is produced either by burning Sulphur or by its release from the compressed liquefied form. Metabisulphite are more stable to oxidation than bisulphites, which in turn show greater stability than sulphites. The antimicrobial action of sulphur dioxide against yeasts, moulds and bacteria is selective, with some species being more resistant than others.

Sulphur dioxide and sulphites are used in the preservation of a variety of food products. In addition to wines these include dehydrated/dried fruits and vegetables, fruit juices, acid pickles, syrups, semi-processed fruit products, etc. In addition to its antimicrobial effects, sulphur dioxide is added to foods for its antioxidant and reducing properties, and to prevent enzymatic and non-enzymatic browning reactions. Carbon dioxide (CO_2) is a colourless, odourless, non-combustible gas, acidic in odour and flavour. In commercial practice it is sold as a liquid under pressure (58 kg per cm^3) or solidified as dry ice.

Carbon dioxide is used as a solid (dry ice) in many countries as a means of low-temperature storage and transportation of food products. Beside keeping the temperature low, as it sublimes, the gaseous CO_2 inhibits growth of psychrotrophic micro-organisms and prevents spoilage of the food (fruits and vegetables, etc.). Carbon dioxide is used as a direct additive in the storage of fruits and vegetables. In the controlled/ modified environment storage of fruit and vegetables, the correct combination of O_2 and CO_2 delays respiration and ripening as well as retarding mould and yeast growth. The final result is an extended storage of the products for transportation and for consumption during the off-season. The amount of CO_2 (5-10%) is determined by factors such as nature of product, variety, climate and extent of storage.

Chlorine: The various forms of chlorine constitute the most widely used chemical sanitiser in the food industry. These chlorine forms include chlorine

(Cl_2), sodium hypochlorite ($NaOCl$), calcium hypochlorite ($Ca(OCl)_2$) and chlorine dioxide gas (ClO_2). These compounds are used as water adjuncts in processes such as product washing, transport, and cooling of heat-sterilised cans; in sanitising solutions for equipment surfaces, etc. Important applications of chlorine and its compounds include disinfection of drinking water and sanitation of food processing equipment.

Chemical food preservatives have to be used only at a dosage level which is needed for a normal preservation and not more. "Reconditioning" of chemical preserved food, *eg.* a new addition of preservative in order to stop a microbiological deterioration already occurred is not recommended. The use of chemical preservatives MUST be strictly limited to those substances which are recognised as being without harmful effects on human beings' health and are accepted by national and international standards and legislation.

Factors related to the preservatives

Factors related to the chemical preservatives:

a) chemical composition;

b) concentration.

Factors related to micro-organisms:

a) micro-organism species; as a general rule it is possible to take the following facts as a basis:

   — sulphur dioxide and its derivatives can be considered as an "universal" preservative; they have an antiseptic action on bacteria as well as on yeasts and moulds;

   — benzoic acid and its derivatives have a preservative action which is stronger against bacteria than on yeasts and moulds;

   — sorbic acid acts on moulds and certain yeast species; in higher dosage levels it acts also on bacteria, except lactic and acetic ones;

   — formic acid is more active against yeasts and moulds and less on bacteria.

b) the initial number of micro-organisms in the treated product determines the efficiency of the chemical preservative.

The efficiency is less if the product has been contaminated because of preliminary careless hygienic treatment or an incipient alteration. Therefore, with a low initial number of micro-organisms in the product, the preservative dosage level could be reduced.

Specific factors related to the product to be preserved:

a) product chemical composition;

b) influence of the pH value of the product: the efficiency of the majority of chemical preservatives is higher at lower pH values, *ie*. when the medium is more acidic.

c) physical presentation and size which the product is sliced to: the chemical preservative's dispersion in food has an impact on its absorption and diffusion through cell membranes on micro-organisms and this determines the preservation effect.

Therefore, the smaller the slicing of the product, the higher the preservative action. Preservative dispersion is slowed down by viscous foods (concentrated fruit juices, etc.)

Miscellaneous factors

a) *Temperature*: chemical preservative dosage level will be established as a function of product temperature and characteristics of the micro-flora;

b) *Time*: at preservative dosage levels in employed in industrial practice, the time period needed in order to obtain a "chemical sterilisation" is a few weeks for benzoic acid and shorter for sulphurous acid.

Usual accepted chemical food preservatives are detailed in Table 6.

Table 6: Chemical Food Preservatives

Agent Acceptable	*Daily intake (mg/Kg body weight)*	*Commonly used levels (%)*
Lactic acid	No limit	No limit
Citric acid	No limit	No limit
Acetic acid	No limit	No limit
Sodium Diacetate	15	0.3-0.5
Sodium benzoate	5	0.03-0.2
Sodium propionate	10	0.1-0.3
Potassium sorbate	25	0.05-0.2
Methyl paraben	10	0.05-0.1
Sodium nitrite	0.2	0.01-0.02
Sulphur dioxide	0.7	0.005-0.2

Some food products in common usage are:

- *Citric acid:* fruit juices; jams; other sugar preserves;
- *Acetic acid:* vegetable pickles; other vegetable products;
- *Sodium benzoate:* vegetable pickles; preserves; jams; jellies; semi-processed products;
- *Sodium propionate:* fruits; vegetables;
- *Potassium sorbate:* fruits; vegetables; pickled products; jams, jellies;
- *Methyl paraben:* fruit products; pickles; preserves;
- *Sulphur dioxide:* fruit juices; dried/dehydrated fruits and vegetables; semi-processed products.

Acidification

Food acidification is a means of preventing their deterioration in so far as a non-favourable medium for micro-organisms development is created. This acidification can be obtained by two ways: natural acidification and artificial acidification.

Natural acidification

This is achieved by a predominant lactic fermentation which assures the preservation based on acidoceno-anabiosys principle; preservation by lactic fermentation is called also biochemical preservation. Throughout recorded history food has been preserved by fermentation. In spite of the introduction of modern preservation methods, lactic acid fermented vegetables still enjoy a great popularity, mainly because of their nutritional and gastronomic qualities.

The various preservation methods discussed thus far, based on the application of heat, removal of water, cold and other principles, all have the common objective of decreasing the number of living organisms in foods or at least holding them in check against further multiplication. Fermentation processes for preservation purposes, in contrast, encourage the multiplication of micro-organisms and their metabolic activities in foods. But the organisms that are encouraged are from a select group and their metabolic activities and end products are highly desirable. The extent of this desirability is emphasised by a partial list of fermented fruits and vegetable products from various parts of the world in Table 7.

Table 7: Some industrial fermentation processes in food industries

I. Lactic acid bacteria	
- cucumbers	dill pickles, sour pickles
- cabbage	sauerkraut
- turnips	sauerruben
- lettuce	lettuce kraut
- mixed vegetables, turnips, radish, cabbage	
- mixed Chinese vegetables, cabbage	Kimchi
- vegetables and milk	Tarhana
- vegetables and rice	Sajur asin
II. Lactic acid bacteria with other micro-organisms	
- with yeasts	Nukamiso pickles
- with moulds	tempeh, soy sauce
III. Acetic acid bacteria—wine, cider or any alcoholic and sugary or starchy products may be converted to vinegar	
IV. Yeasts	
- fruit	wine, vermouth

There are some characteristic features in the production of fermented vegetables which will be pointed out below using cucumbers as an example. In the production of lactic acid fermented cucumbers, the raw material is put into a brine without previous heating. Through the effect of salt and oxygen deficiency the cucumber tissues gradually die. At the same time, the semi-permeability of the cell membranes is lost, whereby soluble cell components diffuse into the brine and serve as food substrate for the micro-organisms.

Under such specific conditions of the brine the lactic acid bacteria succeed in overcoming the accompanying micro-organisms and lactic acid as the main metabolic products is formed. Under favourable conditions (for example moderate salt in the brine, use of starter cultures) it takes at least 3 days until the critical pH value of 4.1 or less—desired for microbiological reasons—is reached. Beside the typical taste, for the consumer a crisp texture is the most important quality criterion for fermented vegetables.

Because there is no heating step before the fermentation, the indigenous plant enzymes in the fermenting materials are still present during the very first phase. After the destruction of the cell membranes they easily get to their active sites and under favourable conditions they can easily cause softening. The environmental conditions act in a different manner on single enzymes or enzymes systems: some enzymes are strongly inhibited by salt, others are activated, and in the acid pH-region many enzymes are irreversibly inactivated. Beside indigenous enzymes also enzymes produced by micro-organisms can be responsible for the undesired soft products.

In technically advanced societies the major importance of fermented foods has come to be variety they add to the diet. However, in many less developed areas of the world, fermentation and natural drying are the major food preservation methods and as such are vital to survival of a large proportion of the world's current population. Artificial acidification is carried out by adding acetic acid which is the only organic acid harmless for human health and stable in specific working conditions; in this case biological principles of the preservation are acidoanabiosys and, to a lesser extent, acidoabiosys.

Combined acidification is a preservation technology which involves as a preliminary processing step a weak lactic fermentation followed by acidification (vinegar addition). The two main classes of vegetables preserved by acidification are sauerkraut and pickles; the definitions of these products adapted from US Code of Federal Register are as follows.

— *Bulk sauerkraut*: Bulk or barrelled sauerkraut is the product of characteristic acid flavour, obtained by the full fermentation, chiefly lactic, of properly prepared and shredded cabbage in the presence of 2-3% salt. On completion of fermentation, it contains not more than 1.5% of acid, expressed as lactic acid.

— *Canned sauerkraut*: Canned (or packaged) sauerkraut, is prepared from clean, sound, well-matured heads of the cabbage plant (Brassica oleracea var. capitata L.) which have been properly trimmed and cut; to which salt is added and which is cured by natural fermentation.

The product may or may not be packed with pickled peppers, pimientos, or tomatoes or contain other flavouring ingredients to give the product specific flavour characteristics. The product

a) may be canned by processing sufficiently by heat to assure preservation in hermetically sealed containers; or

b) may be packaged in sealed containers and preserved with or without the addition of benzoate of soda or any other ingredient permissible under the provisions of Food and Drug Administration (FDA).

Pickles: "Pickles" means the product prepared entirely or predominantly from cucumbers (Cucumis sativus L.). Clean, sound ingredients are used which may or may not have been previously subjected to fermentation and curing in a salt brine (solution of sodium chloride, NaCl). The prepared pickles are packed in a vinegar solution to which may be added salt and other vegetables, nutritive sweeteners, seasonings, flavourings, spices, and other ingredients permissible under FDA regulations. The product is packed in suitable containers and heat treated, or otherwise processed to assure preservation.

Sauerkraut and pickle products can be preserved under the effect of natural or added acidity, followed by pasteurisation when this acidification is not sufficient. Sauerkraut is a very good source of vitamin C; the importance of this product should be emphasised in developing countries as a simple technology which can be applied mainly for consumption of the finished products in remote, isolated areas during the cold season. It is also a excellent technology to be learned to schools which have their own source of cabbage and cucumbers through school agricultural farms.

Sauerkraut and pickles are manufactured on an industrial scale in significant quantities world-wide. However, the basic technology is simple and could be applied at home, farm and community level after some explanation and training. The natural acidification preservation could be considered similar to sun/solar drying in terms of training and development.

Preservation with Sugar

The principle of this technology is to add sugar in a quantity that is necessary to augment the osmotic pressure of the product's liquid phase at a level which will prevent microorganism development. From a practical point of view, however, it is usual to partially remove water (by boiling) from the product to be preserved, with the objective of obtaining a higher sugar concentration. In concentrations of 60% in the finished products, the sugar generally assures food preservation.

It is important to know the ratio between the total sugar quantity in the finished product and the total sugar concentration in the liquid phase because this determines, in practice, the sugar preserving action. The percent composition of a product preserved with sugar, for example marmalade, can be expressed as follows:

$$[i + S + s + n + w] = 100;$$

i = insoluble substance;

s = sugar from fruits;

S = added sucrose;

n = soluble "non sugar"

w = water.

In this case, total sugar concentration, in the liquid phase, of the finished product is:

$$X = 100\ (S + s)\ /\ 100 - (n + i)\ [\%]$$

Therefore, in the case of a standard marmalade with 55 % sugar added (calculated on the finished product basis), the real concentration in the liquid phase is for example:

$$X = 100\ (55 + 8)\ /\ 100 - (5 + 3) = 68.5\%$$

In the food preservation with sugar, the water activity cannot be reduced below 0.845; this value is sufficient for bacteria and neosmophile yeast inhibition but does not prevent mould attack. For this reason, various means are used to avoid mould development:

- finished product pasteurisation (jams, jellies, etc.);
- use of chemical preservatives in order to obtain the antiseptisation of the product surface.

It is very important from a practical point of view to avoid any product contamination after boiling and to assure an hygienic operation of the whole technological process (this will contribute to the prevention of product moulding or fermentation). Storage of the finished products in good conditions can only be achieved by ensuring the above level of water activity.

Heat Preservation

There are various degrees of preservation by heating; a few terms have to be identified and understood.

a) *Sterilisation*: By sterilisation we mean complete destruction of micro-organisms. Because of the resistance of certain bacterial spores to heat, this frequently means a treatment of at least 121°C (250°F) of wet heat for 15 minutes or its equivalent. It also means that every particle of the food must receive this heat treatment. If a can of food is to be sterilised, then immersing it into a 121°C pressure cooker or retort for the 15 minutes will not be sufficient because of relatively slow rate of heat transfer through the food in the can to the most distant point.

b) *"Commercially sterile"*: Term describes the condition that exists in most of canned or bottled products manufactured under Good Manufacturing Practices procedures and methods; these products generally have a shelf-life of two years or more.

c) Pasteurised means a comparatively low order of heat treatment, generally at a temperature below the boiling point of water. The more general objective of pasteurisation is to extend product shelf-life from a microbial and enzymatic point of view; this is the objective when fruit or vegetable juices and certain other foods are pasteurised. Pasteurisation is frequently combined with another means of preservation—concentration, chemical, acidification, etc.

d) Blanching is a type of pasteurisation usually applied to vegetables mainly to inactivate natural food enzymes. Depending on its severity, blanching will also destroy some microorganisms.

Since heat sufficient to destroy micro-organisms and food enzymes also usually has adverse effects on other properties of foods, in practice the minimum possible heat treatment should be used which can guarantee freedom from pathogens and toxins and give the desired storage life; these aims will determine the choice of heat treatment.

In order to safely preserve foods using heat treatment, the following must be known:

a) what time-temperature combination is required to inactivate the most heat resistant pathogens and spoilage organisms in one particular food?

b) what are the heat penetration characteristics in one particular food, including the can or container of choice if it is packaged?

Preservation processes must provide the heat treatment which will ensure that the remotest particle of food in a batch or within a container will reach a sufficient temperature, for a sufficient time, to inactivate both the most resistant pathogen and the most resistant spoilage organisms if it is to achieve sterility or "commercial sterility", and to inactivate the most heat resistant pathogen if pasteurisation for public health purposes is the goal. Different foods will support growth of different pathogens and different spoilage organisms so the target will vary depending upon the food to be heated. Food acidity/pH value has a tremendous impact on the target in heat preservation/ processing.

In a simplified manner, the main operations employed in heat preservation can be described as follows:

— *Food preparation*: Preparation procedures will vary with the type of food. For fruit, washing, sorting, grading, peeling, cutting to size, pre-cooking and pulping operations may be employed.

— *Can/receptacle*: This may be carried out manually or by using sophisticated filling machinery. The ratio of liquid to solid in the can must be carefully controlled and the can must not be overfilled. A headspace of 6-9 mm depth (6-8% of the container volume) above the level of food in the can is usual.

— *Vacuum production*: This can be achieved by filling the heated product into the can, by heating the can and contents after filling, by evacuating the headspace gas in a vacuum chamber, or by injecting superheated steam into the headspace. In each case the can end is seamed on immediately afterwards.

— *Thermal processing*: The filled sealed can must be heated to a high temperature for a sufficient length of time to ensure the destruction of spoilage micro-organisms. This is usually carried out in an autoclave or retort, in an environment of steam under pressure.

— *Cooling*: The processed cans must be cooled in chlorinated water to a temperature of 37°C. At this temperature the heat remaining is sufficient to allow the water droplets on the can to evaporate before labelling and packing.

— *Labelling and packing*: Labels are applied to the can body, and the cans are then packed into cases.

In principle, all these operations can also be carried out at the farm/ community level using the appropriate small scale equipment, preferably only glass jars.

Pasteurisation Process

Physical and chemical factors which influence pasteurisation process are the following:

a) temperature and time;

b) acidity of the products;

c) air remaining in containers.

In pasteurising certain acid juices for example, there are two categories of processes:

a) Low pasteurisation where pasteurisation time is in the order of minutes and related to the temperature used; two typical temperature/time combinations are as following:

   63°C to 65°C over 30 minutes or

   75°C over 8 to 10 minutes.

Pasteurisation temperature and time will vary according to:

— nature of product; initial degree of contamination;

— pasteurised product storage conditions and shelf life required.

In this first category of pasteurisation processes it is possible to define three phases:

— heating to a fixed temperature;

— maintaining this temperature over the established time period (= pasteurisation time);

— cooling the pasteurised products: natural (slow) or forced cooling.

b) Rapid, high or flash pasteurisation is characterised by a pasteurisation time in the order of seconds and temperatures of about 85° to 90°C or more, depending on holding time. Typical temperature/time combinations are as follows:

   — 88°C (190°F) for 1 minute;

   — 100°C for 12 seconds;

   — 121°C for 2 seconds.

While bacterial destruction is very nearly equivalent in low and in high pasteurisation processes, the 121°C/2 seconds treatment give the best quality products in respect of flavour and vitamin retention. Such short holding times, however, require special equipment which is more difficult to design and generally is more expensive than the 63-65°C/30 minutes type of processing equipment. In flash pasteurisation the product is heated up rapidly to pasteurisation temperature, maintained at this temperature for the required time, then rapidly cooled down to the temperature for filling, which will be performed in aseptic conditions in sterile receptacles.

Taking into account the short time and rapid performance of this operation, flash pasteurisation can only be achieved in continuous process, using heat exchangers. Industrial applications of pasteurisation process are mainly used as a means of preservation for fruits and vegetable juices and specially for tomato juice.

Thermopenetration: The thermopenetration problem is extremely important, especially in the case of the pasteurisation of products packed in glass containers because it is the determining factor for the success of the whole operation. During pasteurisation it is necessary that a sufficient heat quantity is transferred through the receptacle walls; this is in order that the product temperature rises sufficiently to be lethal to micro-organisms throughout the product mass.

The most suitable and practical method to speed up thermopenetration is the movement of receptacles during the pasteurisation process. Rapid rotation of receptacles around their axis is an efficient means to accelerate heat transfer, because this has the effect, among others of rapidly mixing the contents. The critical speed of for this movement is generally about 70 rotations per minute (RPM). This enables a more uniform heating of products, reducing heating time and organoleptic degradation.

Heating may precede or follow packaging. These principles of different temperature time combinations very largely determine the design parameters for heat preservation equipment and commercial practices. The food processor will employ no less than that heat treatment which gives the necessary degree of micro-organism destruction. This is further ensured by periodic inspection by local sanitary authorities or by the importing countries sanitary services. However, the food processor also will want to use the mildest effective heat treatment to ensure highest food quality.

It is convenient to separate heat preservation practices into two broad categories: one involves heating of foods in their final containers, the other employs heat prior to packaging. The latter category includes methods that are inherently less damaging to food quality, where the food can be readily subdivided (such as liquids) for rapid heat exchange. However, these methods then require packaging under aseptic or nearly aseptic conditions to prevent or at least minimise recontamination.

On the other hand, heating within the package frequently is less costly and produces quite acceptable quality with the majority of foods and most of our present canned food supply is heated in the package. In practice, therefore, most of the canned food produced locally in developing countries should be heated within the package.

Food Irradiation

Food irradiation is one of the food processing technologies available to the food industry to control organisms that cause food-borne diseases and to reduce food losses due to spoilage and deterioration. Food irradiation technology offers some advantages over conventional processes. Each application should be evaluated on its own merit as to whether irradiation provides a technical and economical solution that is better than traditional processing methods.

Table 8: Possible causes of spoilage (real or apparent) in canned goods
Type of food: Acid and high acid foods (canned fruits)

Condition of can	*Action to be taken to identify cause*
Insufficient vacuum or headspace	Check vacuum and headspace in relation to storage temperature and altitude
"Springer" or "flipper"	Cool can to 15°C and check if still domed. Check can for denting, if possible measure headspace volume change brought about by dents, by comparing can volume with volume of a sound can
Hydrogen swell	Check degree of detinning in can especially at the liquid level. Look for scratches or pinholes in lacquer or tin coating. Check if can is still domed on cooling to 15°C.
"Hard" or "soft swell"	Leaker spoilage. Check can for gross seam faults, perforation due to corrosion or damage to seams. Examine contents for signs of spoilage and can interior for detinning at air/product interface.

For products where irradiation is permitted, commercial applications depend on a number of factors including the demand for the benefits provided, competitiveness with alternative processes and the willingness of consumers to buy irradiated food products. There are a number of applications of food irradiation. For each application it is important to determine the optimum dosage range required to achieve the desired effect. Too high a dosage can produce undesirable changes in texture, colour and taste of foods.

Shelf-life extension: Irradiation can extend the shelf-life of foods in a number of ways. By reducing the number of spoilage organisms (bacteria, mould, fungi), irradiation can lengthen the shelf life of fruits and vegetables. Since ionising radiation interferes with cell division, it can be used as an alternative to chemicals to inhibit sprouting and thereby extend the shelf life of potatoes, onions and garlic. Exposure of fruits and vegetables to ionising radiation slows their rate of ripening. Strawberries, for example, have been found to be suitable for irradiation. Their shelf-life can be extended three-fold, from 5 to 15 days.

Disinfestation. Ionising radiation can also be used as an alternative to chemical fumigants for disinfestation of grains, spices, fruits and vegetables. Many countries prohibit the importation of products suspected of being contaminated with live insects to protect the importing country's agricultural base. With the banning of certain chemical fumigants, irradiation has the potential to facilitate the international shipment of food products.

Packaging Materials

The following are among the more important general requirements and functions of food packaging materials/ containers:

a) they must be non-toxic and compatible with the specific foods;
b) sanitary protection;
c) moisture and fat protection;
d) gas and odour protection;
e) light protection;
f) resistance to impact;
g) transparency;
h) tamperproofness;
i) ease of opening;
j) pouring features;
k) reseal features;
l) ease of disposal;
m) size, shape, weight limitations;
n) appearance, printability;

o) low cost;

p) special features.

Containers

The terms primary and secondary containers have been used. Some foods are provided with efficient primary containers by nature, such as nuts, oranges, eggs and the like. In packaging these, we generally need only a secondary outer box, wrap, or drum to hold units together and give gross protection. Other foods such as milk, dried eggs and fruit concentrates often will be filled into primary containers such as plastic liners which are then packaged within protective cartons or drums. In this case the secondary container provided by the carton or drum greatly minimises the requirements that must be met by the primary container.

Except in special instances, secondary containers are not designed to be highly impervious to water vapour and other gases, especially at zones of sealing, dependence for this being placed upon the primary container. Since primary containers by definition are those which come in direct contact with the food, we will be far more concerned with them than with secondary containers.

Hermetic Closure

Two conditions of the greatest significance in packaging are hermetic and non-hermetic closure. The term hermetic means a container which is absolutely impermeable to gases and vapours throughout its entirety, including its seams. Such a container, as long as it remains intact, will automatically be impervious to bacteria, yeasts, moulds, and dirt from dust and other sources since all of these agents are considerably larger than gas or water vapour molecules.

On the other hand, a container which prevents entry of micro-organisms, in many instances will be non-hermetic. A container that is hermetic not only will protect the product from moisture gain or loss, and from oxygen pickup from the atmosphere, but is essential for strict vacuum and pressure packaging. The most common hermetic containers are rigid metal cans and glass bottles, although faulty closures can make them non-hermetic. With very rare exceptions flexible packages are not truly hermetic for one or more of the following reasons.

First, the thin flexible films, even when they do not contain minute pinholes, generally are not completely gas and water-vapour impermeable although the rates of gas and water vapour transfer may be exceptionally slow; second, the seals are generally good but imperfect; and third, even where film materials may be gas- and water-vapour-tight, such as certain gages of aluminium foil, flexing of packages and pouches leads to minute pinholes and crease holes.

Hermetic rigid aluminium containers can be readily formed without side seams or bottom end seams. The only seam then to make hermetic is the top end double seam, which may be closed on regular tin can sealing equipment. Glass containers are hermetic provided the lids are tight. Lids will have inside rings of plastic or cork. Many glass containers are vacuum packed and the tightness of the cover will be augmented by the differential of atmospheric pressure pushing down the cover. Crimping of the covers, as in the case of pop bottle caps which operate against positive internal pressure, also can make a gas-tight hermetic seal. But bottles fail more often than cans in becoming non-hermetic.

Films Foils and Plastics

Films and foils have different values for moisture and gas permeability, strength, elasticity, inflammability and resistance to insect penetration and many of these characteristics depend upon the film's thickness. Important characteristics of the types of films and foils commonly used in food packaging are given in Table 9. For the most part such films are used in the construction of inner containers. Since they are non-rigid, their main functions are to contain the product and protect it from contact with air or water vapour. Their capacity to protect against mechanical damage is limited, particularly when thin films are considered.

These materials can exist in many forms, depending upon such variables as identity and mixture of polymers, degree of polymerisation and molecular weight, spatial polymer orientation, use of plasticisers (softeners) and other chemicals, methods of forming such as casting, extrusion or calendering, etc. One of the newer classes of plastic materials referred to as copolymers illustrate what can be done with mixtures of the basic units from which plastics are built. The term copolymer refers to a mixture of chemical species in the resin from which films and other forms can be made. The many variations possible make copolymers an important class of plastics to extend the range of useful food packaging applications.

Table 9: Properties of packaging films

Material	*Properties*
Paper	Strength; rigidity; opacity; printability.
Aluminium foil	Negligible permeability to water-vapour, gases and odours; grease proof, opacity and brilliant appearance; dimensional stability; dead folding characteristics.
Cellulose film (coated)	Strength; attractive appearance; low permeability to water vapour (depending on the type of coating used), gases, odours and greases; printability.
Polythene	Durability; heat-sealability; low permeability to water-vapour; good chemical resistance; good low-temperature performance.
Rubber hydrochloride	Heat-sealability; low permeability to water vapour, gases, odours and greases; chemical resistance.
Cellulose acetate	Strength; rigidity; glossy appearance; printability; dimensional stability.
Vinylidene chloride	Low permeability to water vapour, gases, copolymer odours and greases; chemical resistance; heat-sealability.
Polyvinyl chloride	Resistance to chemicals, oils and greases; heat-sealability.
Polyethylene terephtalate	Strength; durability; dimensional stability; low permeability to gases, odours and greases.

Plastic Sheets

— Cellophane paper can be used for packing of dried products, mainly for dried fruit leathers.

— Polyethylene sheets have a variety of uses. They are flexible, transparent and have a perfect resistance to low temperatures and impermeability to water vapour. An important advantage is that these sheets can be easily heat-sealed. Utilisation is in forms of sheets and bags. It is a good packing material for primary protection of dehydrated products. If a good protection is needed to prevent flavour and gas losses, it will be necessary to combine polyethylene with other materials.

Receptacles

In this class there are three categories:

a) *receptacles that can be heat treated:* boxes, bottles and bags. Sterilisable bags used up to 120°C can be manufactured from same raw materials as described under plastic sheets and up to 100°C from cellophane. Polyethylene bags could be used to some extent for packing and pasteurisation of sauerkraut.

b) receptacles that are not heat treated during processing of fruit and vegetables, also divided in bags and boxes. Bags are the most used type of packing from plastic materials and they are manufactured from polyethylene or cellophane; an important utilisation is for dried/ dehydrated fruits and vegetables.

c) special packagings—which can be contacted by action of heat once the finished product is already inside the pack and the air is evacuated.

Laminates

Various flexible materials such as papers, plastic films, and thin metal foils have different properties with respect to water vapour transmission, oxygen permeability, light transmission, burst strength, pin holes and crease hole sensitivity, etc. and so multi-layers or laminates of these materials which combine the best features of each are used. Commercial laminates containing up to as many as eight different layers are commonly custom designed for a particular product.

Laminations of different materials may be formed by various processes including bonding with a wet adhesive, dry bonding of layers with a thermoplastic adhesive, hot melt laminating where one or both layers exhibit thermoplastic properties, and special extrusion techniques. Such structured plastic films may be complete in themselves or be further bonded to papers or metal foils to produce more complex laminates.

Glass Containers

As far as food packaging is concerned, glass is chemically inert, although the usual problems of corrosion and reactivity of metal closures will of course apply. The principal limitation of glass is its susceptibility to breakage, which may be from internal pressure, impact, or thermal shock, all of which can be greatly minimised by proper matching of the container to its intended use and intelligent handling practices. Here consultation with the manufacturer cannot be over-stressed.

The heavier a jar or bottle for a given volume capacity the less likely it is to break from internal pressure. The heavier jar, however, is more susceptible to both thermal shock and impact breakage. Greater thermal shock breakage of the heavier jar is due to wider temperature differences which cause uneven stress between the outer and inner surfaces of the thicker glass. Greater impact breakage susceptibility of the heavier jar is due to the lower resilience of its thicker wall.

Coatings of various types can markedly reduce each of these types of breakage. These coatings, commonly of special waxes and silicones, lubricate the outside of the glass. As a result, impact breakage is lessened by bottles and jars glancing off one another rather than sustaining direct hits when they are in contact in high speed filling lines. Surface coating after annealing protects glass surfaces from many of the minute scratches appearing in normal handling after annealing ovens; surface coating also improves the high gloss appearance of glass containers and is said to decrease the noise from glass to glass contact at filling lines.

With regard to thermal shock, it is good practice to minimise temperature differences between the inside and outside of glass containers whenever possible. Some manufacturers will recommend that a temperature difference of 44°C (80°F) between the inside and the outside not be exceeded. This requires slow warming of bottles before use for a hot fill and partial cooling before such containers are placed under refrigeration.

Glass used for receptacles in fruit and vegetable processing is a carefully controlled mixture of sand, soda ash, limestone and other materials made molten by heating to about 1500°C (2800°F). Main classes of glass receptacles are:

a) jars which are resistant to heat treatments,

b) jars, glasses, etc. for products not submitted to heat treatment (marmalades, acidified vegetables, etc.);

c) glass bottles for pasteurised products (tomato juice, fruit juices, etc.) or not pasteurised (syrups) and

d) receptacles with higher capacity (flasks, etc.)

Jars for sterilised/pasteurised canned products

These receptacles may replace metal cans, taking into considering both the advantages and disadvantages they present. Advantages are: they do not react to food content; they are transparent and can be manufactured in various shapes; they use cheap raw materials and are reusable. Disadvantages: heavier than metal can of same capacity; fragile; lower thermal conductance and a limited resistance to thermal shocks.

Receptacles in this category must assure a perfect hermeticity after their pasteurisation/sterilisation and cooling and this has to be achieved by the use of metallic (or glass) caps and specific materials for tightness. Taking into account the receptacles' closure method, there are two categories of receptacles:

a) glass jars with mechanical closure;

b) glass jars with pneumatic closure;

Jars for products without heat treatment

For marmalades, jellies and jams glass jars with non hermetic closures using metal, glass or rigid plastic caps are used; however for these products the receptacles mentioned above may also be used. The use of jars with pneumatic closure presents the advantage that some products (*eg.* marmalades, jams) can be filled hot and therefore sterile in receptacles. Pneumatic closing generally protects against negative air action which is in this case eliminated from receptacles.

Glass bottles

These receptacles are widely used both for

a) finished products which need pasteurisation (ea. tomato juice, Knit juices, etc.) and for

b) those which are preserved as such (ea. fruit syrups).

Glass bottles in category a) are closed hermetically with metallic caps, provided with special materials for tightness. For glass bottles in category b) various corks, and aluminium caps with tightness materials may be used.

Glass receptacles with high capacity

In this category there are glass flasks with 3 and 10 litre capacity which can be hermetically closed by a SKO caps system and are resistant to product pasteurisation (ea. tomato juice). As bigger receptacles it is possible to use glass demijohns with usual capacity of 25 and 50 litre; these receptacles are used for preservation of fruit juices by warm process. Closing is performed with flexible rubber hoods.

Paper Packaging

As primary containers few paper products are not treated, coated or laminated to improve their protective properties. Paper from wood pulp and reprocessed waste paper will be bleached and coated or impregnated with such materials as waxes, resins, lacquers, plastics, and laminations of aluminium to improve water vapour and gas impermeability, flexibility, tear resistance, burst strength, wet strength, grease resistance, sealability, appearance, printability, etc.

Paper sheets

— Kraft paper is the brown unbleached heavy duty paper commonly used for bags and for wrapping; it is seldom used as a primary container;

— parchment paper: acid treatment of paper pulp modifies the cellulose and gives water and oil resistance and considerable wet strength to this type of packaging material;

— glassine-type papers are characterised by long wood pulp fibres which impart increased physical strength;

— paper with plastic material sheets.

"Tin Can"/Tinplate

The "tin can" is a container made of tinplate. Tinplate, a rigid and impervious material, consists of a thin sheet of low carbon steel coated on both sides with a very thin layer of tin. It can be produced by dipping sheets of mild steel in molten tin (hot-dipped tinplate) or by the electro-deposition of tin on the steel sheet (electrolytic tinplate). With the latter process it is possible to produce tinplate with a heavier coating of tin on one surface than the other (differentially coated).

Tin is not completely resistant to corrosion but its rate of reaction with many food materials is considerably slower than that of steel. The effectiveness of a tin coating depends on:

a) its thickness which may vary from about 0.5 to 2.0 μm;

b) the uniformity of this thickness;

c) the method of applying the tin which today primarily involves electrolytic plating;

d) the composition of the underlying steel base plate;

e) the type of food, and

f) other factors.

Some canned vegetables including tomato products actually owe their characteristics flavours to a small amount of dissolved tin, without which these products would have an unfamiliar taste. On the other hand, where tin reacts unfavourably with a particular food the tin itself may be lacquer coated. The thickness of tinplate sheets may vary from 0.14 mm to 0.49 mm and is determined by weighing a sheet of known area and calculating the average thickness.

Tinplate sheets may be lacquered after fabrication to provide an internal or external coating to protect the metal surface from corrosion by the atmosphere or through reaction with the can contents. They may also be printed by lithography to provide suitable instructions or information on containers fabricated from tinplate sheets. Under normal conditions the presence of the tin coating provides a considerable degree of electrochemical protection against corrosion, despite the fact that in both types of tinplate the tin coating is discontinuous and minute areas of steel base plate are exposed. With prolonged exposure to humid conditions, however, corrosion may become a serious problem.

The coatings not only protect the metal from corrosion by food constituents but also protect the foods from metal contamination, which can produce a host colour and flavour reactions depending upon the specific food. Particularly common are dark coloured sulphides of iron and tin produced in low acid foods that liberate sulphur compounds when heat processed, and bleaching of red plant pigments in contact with unprotected steel, tin, and aluminium.

REFERENCES

Chan, H.T. Jr. (1984). *Handbook of Tropical Foods*. New York/Basel: Marcel Dekker.

Colin, D. (1992). Recent trends in fruit and vegetable processing. In *Food Science and Technology Today*, 7, (2)), pp. 111 - 116.

Desrosier, N.W. and Desrosier, J.N. (1977). *Technology of Food Preservation*, 4th Edition, AVI Publishing Co. Westport, Conn.

Stafford, A.E. (1984). "Mango". In *Handbook of Tropical Foods*, pp. 399-432. H.T. Chan, Jr. Editor. New York/Basel: Marcel Dekker

Thompson, A.K. (1989). *Recent advances in post-harvest technology of fresh fruits and vegetables.* Cranfield Institute of Technology, Silsoe College, U.K.

8

Methods of Processing Fish

Fish is a highly perishable food which needs proper handling and preservation if it is to have a long shelf life and retain a desirable quality and nutritional value. The central concern of fish processing is to prevent fish from deteriorating. The most obvious method for preserving the quality of fish is to keep them alive until they are ready for cooking and eating. For thousands of years, China achieved this through the aquaculture of carp. Other methods used to preserve fish and fish products include

— the control of temperature using ice, refrigeration or freezing

— the control of water activity by drying, salting, smoking or freeze-drying

— the physical control of microbial loads through microwave heating or ionizing irradiation

— the chemical control of microbial loads by adding acids

— oxygen deprivation, such as vacuum packing.

Usually more than one of these methods is used. When chilled or frozen fish or fish products are transported by road, rail, sea or air, the cold chain must be maintained. This requires insulated containers or transport vehicles and adequate refrigeration. Modern shipping containers can combine refrigeration with a controlled atmosphere.

Fish processing is also concerned with proper waste management and with adding value to fish products. There is an increasing demand for ready to eat fish products, or products that don't need much preparation.

The term fish processing refers to the processes associated with fish and fish products between the time fish are caught or harvested, and the time the final product is delivered to the customer. Although the term refers specifically to fish, in practice it is extended to cover any aquatic organisms harvested for commercial purposes, whether caught in wild fisheries or harvested from aquaculture or fish farming.

Larger fish processing companies often operate their own fishing fleets or farming operations. The products of the fish industry are usually sold to grocery chains or to intermediaries. Fish are highly perishable. A central concern of fish processing is to prevent fish from deteriorating, and this remains an underlying concern during other processing operations.

Fish processing can be subdivided into fish handling, which is the preliminary processing of raw fish, and the manufacture of fish products. Another natural subdivision is into primary processing involved in the filleting and freezing of fresh fish for onward distribution to fresh fish retail and catering outlets, and the secondary processing that produces chilled, frozen and canned products for the retail and catering trades.

Importance of Fish Processing

The fish processing industry is very widespread and quite varied in terms of types of operation, scales of production and outputs. The species of fish processed include cod, tuna, herring, mackerel, pollock, hake, haddock, salmon, anchovy and pilchards. Marine fish account for more than 90% of fish production, with the remainder being fresh water fish and fish produced by aquaculture. In general, fish processing operations are located close to commercial fishing areas. However in some cases catches may be transported long distances or exported for processing.

The end products from fish processing may be fresh, frozen or marinated fillets, canned fish, fish meal, fish oil or fish protein products, such as surimi. Surimi is an important fish product, with the majority of catches for some species used solely for its production. Fish processing most commonly takes place at on-shore processing facilities. However some processing can take place at sea, on board fishing vessels-for example the gutting of oily fish. In some regions of the world, where large sea fleets operate, processing can also take place on board fishing vessels.

For some sea fleets, 100% utilisation of the catch may be required by legislation. This means that the entire processing operation, including fish meal and oil production for offal and fish waste, takes place on board the fishing vessels. Some sectors of the industry are very seasonal. Salmon processing, for example, may operate fewer than 100 days per year during the salmon harvesting season. During this time, plants operate at full capacity with little opportunity for down time and little incentive for waste reduction. The processing of seafoods such as squid, cuttlefish, octopus and mussels has not been included because production of these species is relatively small compared to the fish filleting industry.

Before 1970s, processing of fishery products depended on very simple methods, primarily on the knifes and salt in very small scale, and therefore, fishermen had to go fishing with salt on the boats. As a result, the qualities of the catches were usually not good. It often happened that the seafood had spoiled before reaching consumers in cities due to backward storage facilities. The country's reform policies, starting at the end of 1970s, have vitalised the seafood processing industry. Especially, after 1980s when the market prices of seafood were totally liberalised, the great changes have taken place.

Instead of rough processing, the modern processing techniques have been applied to various fish products. The former salted products now have mainly been replaced by frozen or cold stored products; the big package by small package; and the glass container by the soft tin container. In recent years, domestic cold storage has developed rapidly in terms of number of units and scale.

A multi-variety processing system has been formulated in which fish, shrimp, mussels and algae are the main stored products. Activities of seafood processing include cold storing, freezing, drying, smoking, canning etc. The major types of processed products are categorised as follows:

- frozen products
- dried products
- smoked products
- canned products
- surimi products
- fish meal
- products made by fish oil and fish liver
- additives and seasoning products

— medicines

— alga glue

— art craft (pearl)

Although the sector has gained a great development, there is still a long way to go to meet the increasing demands as at present only about 30% of the total seafood are processed, whereas most of them are still sold on the markets directly from the farms, especially fresh water fish. Among all the processed seafood, most of the share is developed into frozen products, and then in turn dried, smoked, surimi, canned etc. The major constrains facing to the sector now are:

— weak infrastructures

— lack of funds

— backward management

— consumer's habit, most consumers, especially in the south, like live seafood.

As rising demand has led to higher prices, both quality and standards have improved. Differences in prices for different species, standards and quality grades have increased, giving producers and distributors a much better incentive for supplying high quality seafood. The small-scale fisheries of developing countries are vital because they provide a nutritious food which is often cheaper than meat and therefore available to a larger number of people. In many countries the bulk of fish is sold fresh for local consumption. Processing, where this is done, is either to supply distant markets or to produce a range of products with different flavours and textures.

Fish is an extremely perishable food. For example, most fish become inedible within twelve hours at tropical temperatures. Spoilage begins as soon as the fish dies, and processing should therefore be done quickly to prevent the growth of spoilage bacteria. Fish is a low acid food and is therefore very susceptible to the growth of food poisoning bacteria. This is another reason why it should be processed quickly. Some methods of preservation cause changes to the flavour and texture of the fish which result in a range of different products. These include:

— Cooking (for example, boiling or frying)

— Lowering the moisture content

— Lowering the pH (by fermentation)

Lowering the temperature with the use of ice or refrigeration also preserves the fish, but causes no noticeable changes to the texture and flavour.

Fish provides a good source of high quality protein and contains many vitamins and minerals. It may be classed as either white, oily or shellfish. White fish, such as haddock and seer, contain very little fat (usually less than 1%) whereas oily fish, such as sardines, contain between 10-25%. The latter, as a result of its high fat content, contain a range of fat-soluble vitamins (A, D, E and K) and essential fatty acids, all of which are vital for the healthy functioning of the body. The table illustrates some of the main nutritional differences between oily and white fish.

Average composition of fish

Composition	*White fish eg. haddock*	*Oily fish eg. herring*
Energy (KJ)	321	970
Protein (g)	17	17
Fat(g)	07	18
Water (g)	82	64
Calcium (mg)	16	33
Iron (mg)	0.3	08
Vitamin A (m a)	0	45
Thiamine (mg)	0.07	0

Certain processing techniques such as boiling leach the water-soluble vitamins into the surrounding liquid. If this is thrown away, a great deal of nutritional value is lost.

Fish Products

Cooked Fish

Cooking provides short-term preservation of fish and it is usually a few days before any deterioration becomes noticeable. A range of methods are used for cooking fish but the principle of the process remains the same. The flesh of the fish softens, enzymes become inactivated and the process kills many of the bacteria present on the surface of the fish.

Boiling and poaching both involve cooking the fish in hot water whereas frying uses hot oil. The advantage of these techniques is they are very simple and require no more than basic household equipment and are therefore suitable for small-scale production. Cooked fish products are most usually for immediate consumption and require no sophisticated packaging. The shelf-life can be extended for a few days by using refrigerated storage and the product should be covered to prevent recontamination.

Frozen Fish

The spoilage of fish is directly related to temperature. The higher the temperature, the faster the spoilage up to around 40°C, above which heat will destroy bacteria and enzymes. Any reduction in the temperature prior to processing will maintain the quality of the fish for longer. Fish can be kept cool by covering it with clean, damp sacking and placing it in the shade. Although this method is simple and requires no special equipment, the fish still begins to deteriorate within a few hours. An alternative is to pack the fish with ice. This is an effective method and preserves the fish for a longer period of time. Obtaining ice, however, can be difficult for the following reasons:

— Most ice-making machines are power-operated and therefore require some kind of fuel. Obtaining fuel can often be difficult and the machines may prove expensive to operate.

— A great deal of ice is required and often the cost of the ice is greater than the actual cost of the fish.

Freezing is an alternative method for cooling fish. This technique provides long-term preservation, but it is relatively expensive in terms of equipment and operating costs. In view of this it is not recommended for the majority of small-scale fisheries.

Cured Fish

Curing involves the techniques of drying, dry salting/brining (soaking in salt solution) or smoking. These may be used alone or in various combinations to produce a range of products with a long shelf-life. For example:

— Drying—Smoking—Drying

— Brining—Smoking—Drying

— Salting—Drying

— Salting—Drying—Smoking

Techniques such as these reduce the water content in the flesh of the fish, and thereby prevent the growth of spoilage microorganisms.

Dried Fish

The heat of the sun and movement of air remove moisture which causes the fish to dry. In order to prevent spoilage, the moisture content needs to be reduced to 25 per cent or less. The percentage will depend on the oiliness of the fish and whether it has been salted. Traditionally, whole small fish or split large fish are spread in the sun on the ground, or on mats, nets, roofs, or on raised racks. Sun-drying does not allow very much control over drying times, and it also exposes the fish to attack by insects or vermin and allows contamination by sand and dirt. Such techniques are totally dependent upon the weather conditions. The ideal is dry weather with low humidity and clear skies.

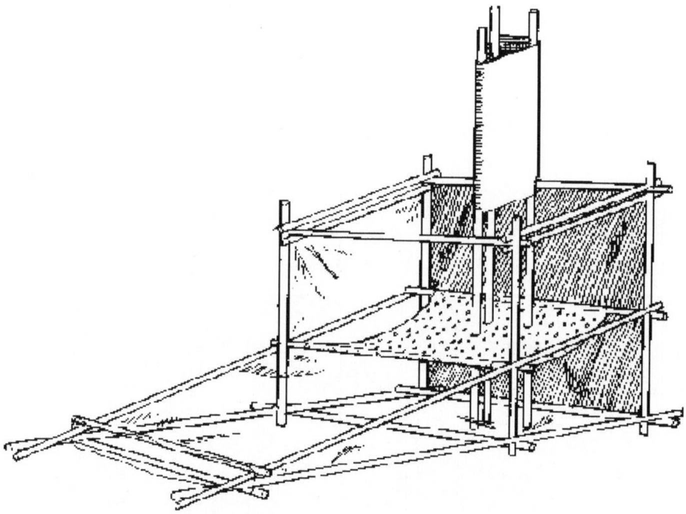

Fig. 1: Improved solar dryer

Alternatives to sun-drying involve the use of solar or artificial dryers. There has been a great deal of research on the development of solar dryers as an improved method of drying fish. This has shown that by achieving increased drying temperatures and reduced humidities, solar dryers can increase drying rates and produce a lower moisture content in the final products, with improvements in fish quality compared with the traditional sun-drying techniques. Fig. 1 illustrates a solar tent dryer. This was first developed in Bangladesh, but there are now numerous variations in different parts of the world. It is probably one of the most simple designs.

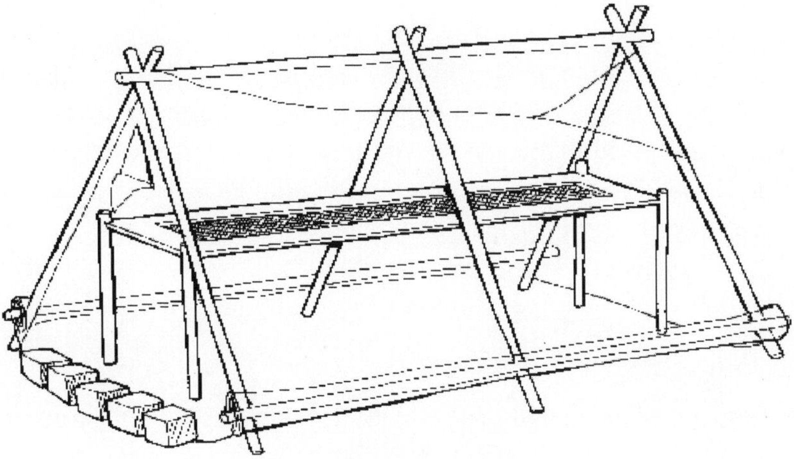

Fig. 2: Solar tent dryer

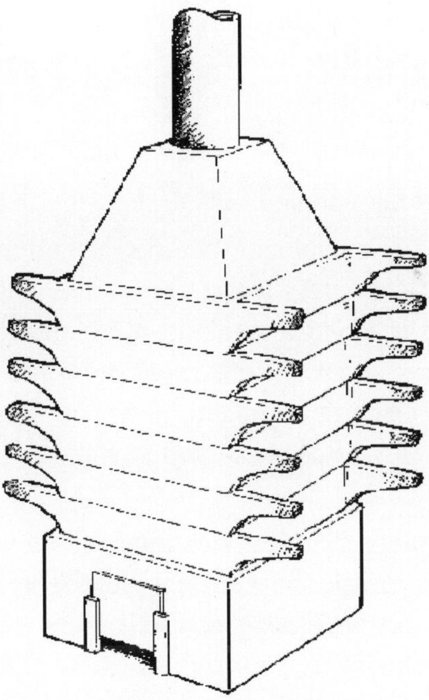

Fig. 3: Tray dryer

Fig. 2 shows an improved solar dryer, with a separate collector and drying chamber. The chimney is painted black to absorb more heat. This will heat the air inside the chimney, thereby increasing the air flow through the dryer.

Fig. 3 illustrates an artificially-heated tray dryer. When rain threatens, the trays, which were previously placed in the sun to dry, are assembled on top of each other over a simple heating compartment. A roof and chimney are placed on top, and drying continues by direct heating. Both solar and artificial dryers try to overcome the difficulties posed by sun-drying during the rainy season. With these dryers it is possible to minimise drying times and to increase the product quality. It should however, be pointed out that it is only advantageous to use such dryers if there is a market for a higher-quality product or if the fish would otherwise be lost.

Salted Fish

Most food poisoning bacteria cannot live in salty conditions and a concentration of 6-10 per cent salt in the fish tissue will prevent their activity. The product is preserved by salting and will have a longer shelf-life. However, a group of micro-organisms known as 'halophilic bacteria' are salt-loving and will spoil the salted fish even at a concentration of 6-10 per cent. Further removal of the water by drying is needed to inhibit these bacteria. During salting or brining two processes take place simultaneously:

— water moves from the fish into the solution outside

— salt moves from the solution outside into the flesh of the fish.

Salting requires minimal equipment, but the method used is important. Salt can be applied in many different ways. Traditional methods involve rubbing salt into the flesh of the fish or making alternate layers of fish and salt (recommended levels of salt usage are 30-40 per cent of the prepared weight of the fish). There is often the problem, however, that the concentration of salt in the flesh is not sufficient to preserve the fish, as it has not been uniformly applied.

A better technique is brining. This involves immersing the fish into a pre-prepared solution of salt (36 per cent salt). The advantage is that the salt concentration can be more easily controlled, and salt penetration is more uniform. Brining is usually used in conjunction with drying. Ultimately the effectiveness of salting for preservation depends upon:

— uniform salt concentration in the fish flesh

— concentration of salt, and time taken for salting

— whether or not salting is combined with other preservation methods such as drying.

Smoked Fish

The preservative effect of the smoking process is due to drying and the deposition in the fish flesh of the natural chemicals of wood smoke. Smoke from the burning wood contains a number of compounds which inhibit bacteria. Heat from the fire causes drying, and if the temperature is high enough, the flesh becomes cooked. Both of these factors prevent bacterial growth and enzyme activity which may cause spoilage. Fish can be smoked in a variety of ways, but as a general principle, the longer it is smoked, the longer its shelf-life will be. Smoking can be categorised as:

— Cold smoking. In this method, the temperature is not high enough to cook the fish. It is not usually higher than 35°C.

— Hot smoking. In this method, the temperature is high enough to cook fish.

Hot smoking is often the preferred method. This is because the process requires less control than cold processing and the shelf-life of the hot-smoked product is longer, because the fish is smoked until dry. Hot smoking does, however, have the disadvantage that it consumes more fuel than the cold-smoking method. Traditionally, the fish would be placed with smouldering grasses or wood. Alternatively, fish may be laid or hung on bamboo racks in the smoke of a fire.

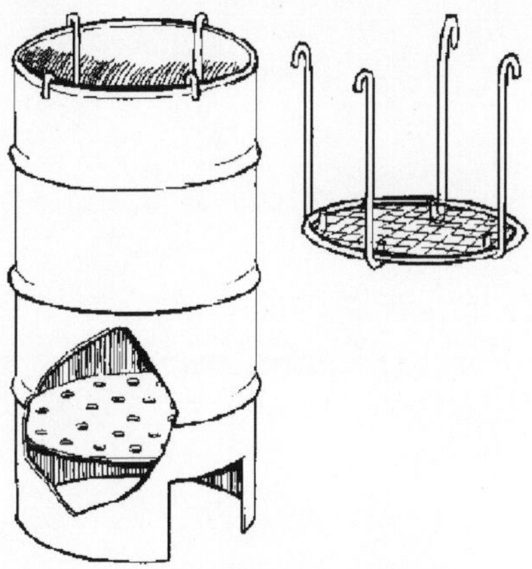

Fig. 4: Oil drum smoker

There are various types of kiln available in different parts of the world, which are used for smoking. Although traditional kilos and ovens have low capital costs, they commonly have an ineffective air-flow system, which results in poor economy of fuelwood and lack of control over temperature and smoke density. Improved smokers include the oil drum smoker and the chorker smoker. As well as improved smokers, there are also improved techniques which involve either pre-salting the fish, so that the moisture content is reduced prior to smoking. Alternatively there are a range of improved kiln and oven designs.

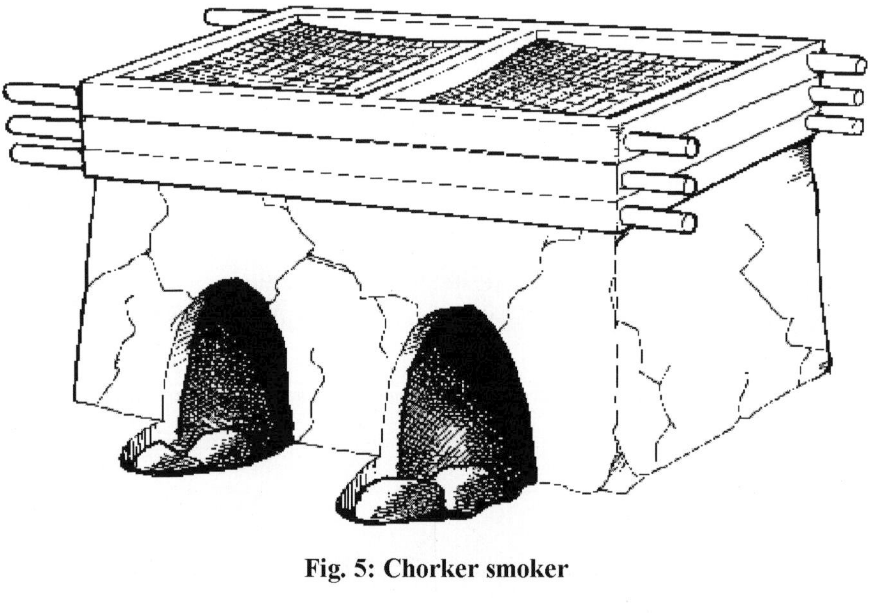

Fig. 5: Chorker smoker

Packaging

The most important concerns regarding packaging for these products are to prevent moisture pick-up and to prevent recontamination by insects and micro-organisms. Traditional packaging materials include cane baskets, leaves, and jute bags. Alternatives include flexible packaging such as polythene bags, or wooden and cardboard packs. Indeed, the two may be combined, as in a polythene bag enclosed in an outer cardboard pack.

Fermented Fish

Fermentation is a process by which beneficial bacteria are encouraged to grow. These bacteria increase the acidity of the fish and therefore prevent the growth of spoilage and food-poisoning bacteria. Additionally, salt is used to prevent the action of spoilage bacteria and allow the fish enzymes and the beneficial acid-producing bacteria to soften (break down) the flesh.

Fermentation is therefore the controlled action of the desirable micro-organisms in order to alter the flavour or texture of the fish and extend the shelf-life. The use of fermentation as a low-cost method of fish preservation is commonly practiced all over the world. There are many different types of fermented products and their nature depends largely on the extent of fermentation which has been allowed to take place. They can be categorised as:

— fish which retains its original texture

— pastes

— liquids/sauces.

As with salting, there is little need for equipment other than pans and containing vessels, and the process may easily be carried out on a small scale.

Fish Paste

This is a product from Eastern Asia. It is made from whole or ground fish, fish roe, or shellfish. It is reddish brown in colour, although this will depend on the raw materials used, and is slightly salty with a cheese-like odour.

Packaging

There are almost as many traditional methods of packaging fermented fish as there are ways of making it—such as earthenware pots, oil cans, drums and glass bottles. In the past, the latter have been used because of their low cost, but nowadays, cheaper plastic containers tend to replace the traditional types. The most important function of packaging for fermented fish products is that the containers should be air-tight, helping to develop and maintain the airless conditions required for good fermentation and storage. As the major advantage of these products is their low cost, the type of packaging is necessarily restricted. Glass bottles are often used for the better-quality products, but earthenware pots and even plastic bags are used.

White Fish Filleting

Filleting involves a number of unit operations: pretreatment, fish filleting, trimming of fillets, packing and storage. These processes generally take place within separate departments of the fish processing plant. White fish species have a low oil content and, unlike their oily fish counterparts, are generally gutted, cleaned and sometimes de-headed on board the fishing vessel. The fish are kept on ice in boxes before being delivered to the fish processing plant. On arrival at the plant, fish may be re-iced and placed in chilled storage until required for further processing.

Pretreatment of the fish involves the removal of ice, washing, grading according to size and de-heading, if this has not been done previously. Large fish may also scaled be before further processing. The next step in the process is filleting, which is generally done by mechanical filleting machines. The filleting department is generally separated from the pretreatment area by a wall, to prevent workers and goods passing from the non-sterile pretreatment area to the sterile filleting area.

The filleting machines comprise pairs of mechanically operated knives which cut the fillets from the backbone and remove the collarbone. Some fish fillets may also be skinned at this stage. In the trimming department, pin bones are removed and operators inspect the fillets, removing defects and any parts that are of inferior quality. Offcuts are collected and minced. Depending on the final product, the fillets may be cut into portions according to weight or divided into parts such as loin, tail and belly flap. As a final step before packaging, the fillets are inspected to ensure they meet product standard.

Fresh products are packaged in boxes with ice, the ice being separated from the products by a layer of plastic. Frozen products can be packed in a number of ways. Fillets or pieces can be individually frozen and wrapped in plastic, but the most common method is for them to be packed as 6-11 kg blocks in waxed cartons. The blocks are typically frozen and then kept in cold storage.

Oily Fish Filleting

Oily fish species are characterised as those having oils distributed throughout the fillet and in the belly cavity around the gut. Fillets from these species may contain up to 30% oil. Oil content varies not only between species but also within species. Oily fish species are very rarely gutted or cleaned on board the fishing vessels, due to the high oil content and the consequent risks associated with oily surfaces. Keeping the skin of the fish intact also reduces oxidation of the oil and thus maintains flesh quality. Oily species can be filleted like white fish species, but they are also used for canning.

CANNING

The fish are off-loaded at the plant, weighed and loaded into water flumes. The flumes transport the fish to holding vessels, where they remain until required for processing. Fish are flumed, as required, onto cutting tables, where the heads, tails and other inedible parts are removed. Some fish species, such as mackerel, need to have the skin removed by immersion in a warm caustic bath. The effluent generated from this process has a high organic load and has to be neutralised before being discharged.

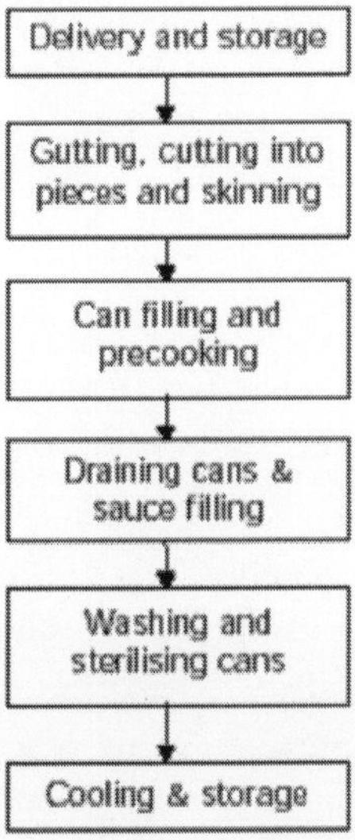

Fig. 6: Process flow diagram of the canning process

The canning process depends on the size of the fish. Small fish species such as sardines and pilchards are generally canned whole, with only the heads and tails removed. These whole-fish products are cooked in the can after it has been filled with brine or oil. Medium-sized fish species are cut into pieces and pre-cooked in the can before the can is filled with brine or oil. For large fish species such as mackerel and tuna, the fish are filleted, cut into pieces of suitable size and also precooked in the can.

Bones and inedible parts are removed when large fish species such as tuna are canned. After precooking, the liquid is drained from the cans and oil, brine or sauces are added. The cans are then sealed, sterilised and then stored. Most large canneries also operate a fish meal plant, in which fish not suitable for canning is combined with offal and processed into fish meal.

Meal and Oil Production

Fish meal and fish oil are produced from fish that are caught specifically for this market, by-catch from fishing activities and solid waste from filleting and canning. Fish meal and fish oil products have a high nutritional value. Fish meal is used as feed for livestock and farmed fish, and the oil is used as an ingredient in paints and margarine. Fish meal is derived from the dry components of the fish, and the oil from the oily component. Water, which makes up the rest of the fish matter, is evaporated during the process.

Most fish meal and fish oil production processes are automated and continuous, and comprise several process lines, each with a certain processing capacity. Production rates vary considerably, according to the season and types of fish being processed. On board the fishing vessels, the catch is normally stored in tanks of water. Upon arriving at the processing plant the fish are pumped to holding bins, where they are stored until required for processing.

Extra sea water may need to be added to pump the fish. From the storage bins, the fish are transported by screw conveyors to a cooking process which acts to coagulates the protein. The cooked mixture is then screened, using a strainer conveyor or a vibrating screen, and then pressed to remove most of the water from the mixture. The pressed cake is shredded and dried, using an indirect steam drier or a direct flame dryer. The meal passes through a vibrating screen and on to a hammer mill, which grinds it to the appropriate size.

The ground meal is automatically weighed and bagged. The pressed liquid generated from the previous processes passes through a decanter to remove most of the sludge, which is fed back to the meal dryer. Oil is separated from the liquid by centrifuges, polished and refined to remove any remaining water and impurities. The separated aqueous phase, referred to as stickwater, is concentrated in an evaporator and then added to the pressed fish meal prior to being sent to the dryer.

Freshwater Fish Processing

The manufacturing potential of the raw material as food depends on two features - the nutritive and the technological value.

Nutritive Value

The nutritive value of dishes prepared from fish and from animal meat is comparable, but in some cases fish-based meals are advisable. In such an evaluation, many parameters, such as energetic value, quality and content of protein components, vitamins and mineral compound content should be examined. The energetic value of eel meat is lower than that of fat beef (1,050 kJ/100 g and 1,250 kJ/100 g respectively); while in the case of trout it amounts to 600 kJ/100 g and it is lower than for lean beef, 735 kJ/100 g. Thus the meat of freshwater fish can be a valuable constituent in low-calorie diets and at the same time has a high energy content.

The composition of amino-acid proteins in fish meat is similar to that of a hen's egg. Consumption of fish together with products of plant origin which are poor in some amino-acids (lysin, threonine), enables not only a complete utilisation of plant protein, but also improves the content of a diet. The biological value of freshwater fish fats is lower than that of marine fish because the former contain fewer unsaturated aliphatic acids. Fish meat is valuable as a source of vitamins and mineral substances. It contains especially the trace metals such as: selenium, molybdenum, cobalt, whose value is emphasised by physiologists.

The definition of food stresses that the basic food ingredients as well as the raw materials used for its production must be wholesome. However, contamination of the environment is fast increasing, especially through the use of chemicals in agriculture or in industry. For that reason, certain countries or groups of countries establish limits and recommendations for permissible levels of chemical contaminants the excess of which leads to exclusion of such raw material from the production of food for human consumption. This problem, for many reasons (diets, habits, analytical methods), is far from being solved as countries have different attitudes in this respect. It may well constitute a non-tariff barrier on a free market in the future.

Technological Value

The technological value generally depends on two parameters: the yield of preliminary processing and the quality features of fish meat and by-products. The yield of edible parts of the fish depends, first, on the species and constitution, and also on age and consequently size and maturity. Yield is affected by the ratio between edible and inedible parts of the fish and this is a decisive factor with regard to the technological value of the fish. This ratio depends on the species.

It is most favourable in the Salmonidae family, amounting to approximately 75% of the weight. For most fish species this parameter ranges from 50 to 60%. In the case of perch and most of the Cyprinidae family the yield is less than 50%. More information on the yield of preliminary processing of freshwater fish is given in Table 1.

Table 1: Average yield of preliminary processing (manual processing) of several species of freshwater fish

Species	*Size of Fish (kg)*	*Form of Preprocessed Fish*	*Yield(%)*
Trout	> 0.35	gutted	74 - 82
Trout	> 0.35	h/g (deheaded/gutted)	62 - 74
Trout	> 0.35	fillet with skin	50 - 55
Carp	> 3.0	gutted	76 - 82
Carp	> 1 - 3.0	gutted	73 - 79
Carp	> 3.0	deheaded and gutted	55 - 61
Carp	> 3.0	chunks	49 - 57
Carp	> 3.0	fillet with skin	41 - 49
Pike-perch	> 1.0	gutted	79 - 89
Pike-perch	> 1.0	deheaded and gutted	66 - 74
Pike-perch	0.35 - 0.5	deheaded and gutted	60 - 68
Pike-perch	> 1.0	chunks	56 - 68
Pike-perch	> 1.0	fillet with skin	52 - 64
Pike	1 - 3.0	gutted	76 - 84
Bream	> 1.0	gutted	68 - 76
Bream	0.5 - 1.0	deheaded and gutted	56 - 64
Bream	0.5 - 1.0	chunks	52 - 64

Evaluation of the technological value of freshwater fish should take account of its possible utilisation for different products, considering the sensory properties such as: flavour, texture, appearance, size and bone content. These parameters are decisive as to consumer's interest and thus the market demand. Fish with high bone content are not so popular as a product for consumption. Therefore, the technological value of roach (Rutilus rutilus) is lower than that of pike-perch (Stizostedion lucioperca).

The taste of freshwater fish depends mainly on the quality of their water habitat and on their food. It is known that fish (for example, carp) living in dirty and muddy ponds, have an unpleasant flavour. The flavour of wild trout from streams is better than that of fish from aquaculture. The opposite is true for eel resulting from the fact that the aquaculture eel has a more tender tissue and thinner skin. Freshwater fish are classified according to size, larger individuals usually being preferred to small fish. This is also connected with bone content: *eg*., trouts weighing about 300 g are very popular as single portions, prices increasing with popularity.

The most expensive are fish weighing over 500 g which are destined for smoking. The best market value are carp of 1-2 kg, but those exceeding 3 kg have less customer appeal. The sanitary and hygienic condition of fish and fish meat also influences the technological value. This relates to the presence of parasites and pathogenic micro-organisms. However, the main role in evaluating technological value and usefulness is played by a set of features termed freshness. These features change during storage after the death of the fish and the intensity of the changes depends on the species, fishing conditions and storage conditions immediately after capture.

Changes in Death Fishes

On the death of the fish, processes of physical and chemical change caused by enzymes and micro-organisms begin to occur. The complete decay of the fish is the final result of those changes. Post-mortem changes which take place in fish tissue occur in the following phases:

— slime secretion on the surface of fish

— rigor mortis

— autolysis as enzymatic decomposition of tissues

— microbiological spoilage

The duration of each phase can change or phases can overlap. This depends on storage conditions, especially the temperature which greatly influences these processes.

Secretion of Slime

Slime is formed in certain cells of fish skin and the process becomes very active just after fish death. Some of the fish, for example eel, secrete more slime than, for comparison, Salmonidae and perch. Fish which secrete great quantities of slime have poorly developed scales; very often the quantity of slime reaches 2-3% of the fish mass and that in turn creates problems during processing. The secretion process stops with the onset of rigor mortis. Slime contains large amounts of nitrogenous compounds and these provide good nourishment for micro-organisms originating from the environment. Therefore, the slime spoils quickly: first giving an unpleasant smell to the fish, and second opening the way for further and deeper bacterial penetration into the fish.

Rigor Mortis

Rigor mortis is a result of complicated biochemical reactions which cause muscle fibres to shorten and tighten, and finally the fish becomes stiff. Rigor mortis has many technological consequences. If, for example, the bones were removed prior to rigor mortis the length of the fillet shortens by 30%. At the same time, the fillet becomes wider and thicker because its volume does not change. This tightness very often causes the connective tissue of individual myomeres to break; this process is termed "gaping" and results in muscle separation which is considered a quality defect.

"Gaping" depends on temperature; the higher the temperature of fish at the beginning of the rigor mortis process the greater the gaping of the muscle. Therefore, during rigor mortis fish temperature should be as low as possible. For example, for roach and perch kept at 0°C rigor mortis begins 24 hours after death and lasts for 72-80 hours. When the same species is kept at 35°C it begins 20-30 minutes after death and stops after about 3 hours.

The time rigor mortis begins and its duration depend on the fish species (*eg*., for carp at 0°C it starts after 48 hours, for roach and perch at 0°C after 24 hours), on the fish catching technique, and on fish temperature. It was also found that fast swimmers, for example trout, undergo rigor mortis faster but for a shorter duration than slow swimmers like carp. In those fish which are in good condition (well-nourished) rigor mortis is more intensive. Fish put to death just after removal from the water reach a state of rigor mortis later than those fish which died after a long agony.

In the case of carp put to death just after capture rigor mortis begins after 48 hours, but if the carp died after a long agony it sets in after 24 hours (at 0°C). Unnecessary and rough handling of the fish can shorten the time of occurrence and duration of rigor mortis. Such treatment causes stress in live fish. Fish body temperature is a decisive factor in the onset and duration of the rigor mortis process.

The higher the temperature the sooner it begins and the faster it ceases. This is evidenced by enzymatic reactions whose speed increases with increased temperature. At high temperatures it results in greater changes in proteins, the latter causing higher loss of tissue juices, *eg*., during processing. Usually, the later rigor mortis begins and the longer it lasts, the longer are the storage life of the fish and its use for consumption.

Autolysis

On the death of the fish, a complicated biochemical process starts, leading to a decomposition of basic compounds of tissues which takes place under the influence of enzymes. This decomposition involves proteins, lipids and carbohydrates. Its intensity is not the same for all compounds and the decomposition of one can influence the decomposition of the others. The quality of fish as a raw material for consumption or for processing depends largely on proteolysis, that is, the decomposition of proteins. This process follows rigor mortis.

The final products of protein hydrolysis, under the influence of enzymes, are: amino-acids and other low-molecular substances which have an impact on the sensory features of fish. A similar situation concerns the products of lipid autolysis: thus autolysis cannot be qualified as a phase in the spoilage process. During autolysis, great changes occur in the structure of muscle tissue which becomes softer, and very often falls into layers along the myosepts.

In small fish, perforation of the belly occurs. From the technological view, it is negative because the proteolysis process leads to a decrease in the capacity of tissue to retain tissue juice, resulting in toughness of texture of the final product. The degradation of proteins creates ideal conditions for the growth of spoilage bacteria.

Microbiological Decomposition

The muscle tissue of live fish is generally sterile but bacteria thrive in the alimentary tract and on the skin, and from there they penetrate into the muscles; for example, through the blood vessels. This process is further favoured by structural changes in the tissue as a result of rigor mortis and autolysis. Bacteria are able to decompose proteins, but products of proteolysis such as amino-acids and other low-molecular nitrogenous compounds provide better nourishment. Thus it was found that, due to lower content of these substances, freshwater fish tissue undergoes microbiological decomposition more slowly than marine fish tissue.

Micro-organisms cause decomposition of not only proteins but other compounds containing nitrogen, lipids to peroxides, aldehydes, ketones and lower aliphatic acids. However, the decomposition of nitrogenous compounds occurs much faster than in the case of lipids. Compounds such ammonia, hydrogen sulphide and mercaptans, indole, skatole, etc., are the final products of microbiological spoilage of fish, which produces an unpleasant and then disgusting flavour. Penetration of bacteria into fish tissue and microbiological decomposition begins with autolysis and these processes are practically parallel. However, their rate and intensity strictly depend on the storage temperature. Low temperature strongly inhibits the activity of micro-organisms in which case the autolysis process dominates.

Fish Freshness Indicators

Freshwater fish, as other fish species, are raw material which fast deteriorates. This implies that both the producer and the consumer are very often exposed to the risk of buying fish which is not fresh or has even deteriorated. Knowledge of the average shelf life for individual fish species - depending on storage conditions - is a basic principle applied in the food - and the fish - industry. Effective, objective and repeatable methods for evaluation of raw material freshness should be specified, but attempts so far are only now showing positive results. Thus, sensory analysis is the main method of evaluating fish freshness.

It enables differences in texture, flavour, and taste to be determined, and subsequently the usefulness of the raw material. Sensory properties change during storage from the desired very high standard, through neutral or average, and finally to undesirable or disgusting. It is generally assumed that prior to disappearance of desirable features the fish is considered to be fresh, while the appearance of undesirable or disgusting features disqualifies the raw material. The most difficult step is to determine an intermediate state in which the fish is not entirely fresh. Sensory analysis is thus carried out on raw fish and cooked fish. Flavour, appearance and state of abdominal cavity (for not eviscerated fish) are the main indicators of quality in the case of raw fish. For cooked fish, smell is the most important indicator.

Role of Packaging

Packaging should protect the product from contamination and prevent it from spoilage, and at the same time it should:

— extend shelf life of a product
— facilitate distribution and display
— give the product greater consumer appeal
— facilitate the display of information on the product

The quality of freshwater fish which is delivered to the consumer or the processing plant as live fish greatly depends on correct handling during transport and, when processed, on suitable packaging. For short distances, the live fish can be transported in insulated containers with lids, capacity varying from 300 to 1,000 kg of fish. Fish can also be transported in normal lorries, but for long distances the water in the containers must be aerated and cooled by portable devices. In order to maintain good quality of fresh fish during transportation, fish boxes made of suitable materials should be used. When purchasing fish boxes the six following requirements should be remembered; they should:

— be of a suitable size for the range of fish to be handled or the product to be put into them
— be of a convenient size for manual handling or lifting by mechanical equipment
— be stackable such that the weight of the containers on top rests on the containers underneath and not on the fish
— be constructed of impervious non-staining materials
— be easy to clean
— provide drainage for melted ice

Fish boxes are usually made of high-density polyethylene. Although this offers many advantages, such as duration, lightness, ease of cleaning, there are also disadvantages, *eg*., high price and the fact that they are not returnable. That is why disposable fish boxes of about 25 kg capacity (fish and ice) are more often used: these include fibreboard cartons, waxed and waterproof boxes. In the case of transport by lorries with no cooling system, insulated cartons, *eg*., boards made of moulded polyestyrene should be preferred.

The latter is commonly used for delivery of chilled and frozen fish and fish products to wholesale and retail outlets. In the case of fillets, each layer of fillets should be packed thin and separated from the ice with a plastic foil. Styropor boxes are normally sold with lids, which fit very closely and can be with or without drainage holes. In a typical range, wall thickness varies with box size; *eg*., a 6 kg capacity box has a 15 mm thick wall, a 10 kg box a 19 mm wall, a 25 kg box a 25 mm wall.

The main disadvantage of moulded polyestyrene fish boxes is their lack of strength. They are easily damaged or broken by rough handling. This limits their size and use. Polyestyrene is difficult to clean. Polyestyrene boxes are difficult to re-use, and are usually non-returnable. They may cause disposal problems due to their bulk. The packaging industry improves its products by using new materials with better insulating properties or by introducing new leakproof designs. The new containers are often lighter and less bulky.

Retail Packaging

The main role of packaging is described above but in respect of retail presentation it should also reduce the smell and the drip, and enable the product to be tucked into shopping baskets with other purchases. Moreover, the packaging of fish products should ensure attractive presentation among other food products without contaminating them. Basic packaging materials include paper, cartons, sheets of metal, metal foils and many kinds of plastics. Despite the rapid growth in use of plastics, the role of paper and carton as packaging materials does not decrease.

Kraft paper or carton are often laminated with polyethylene or aluminium foil which render them waterproof. Such material is used for production of trays for packaging of fresh or frozen products. More often, trays are made of plastic materials such as polyestyrene or expanded polyestyrene. Expanded polyestyrene is frequently used but it is partly oxygen-permeable and so those products which are sensitive to rancidity have to be additionally overwrapped or skin-packed with suitable film.

The materials mentioned above are not stable at high temperatures and hence are not suitable for trays to be used in an oven. Polyester can be used as a packing material for heating of the product in the traditional and microwave ovens, but this material cannot be used for microwave cooking. Trays used for packing are generally overwrapped with a protective film, often with PE wrapping which shrinks. The film shrinking is achieved by use of hot air or hot water. Stretch wrapping is often used for products which are heat-sensitive. The film is stretched over the product manually (very often in the supermarket) or by machine.

Foils used as wrapping or bags for packing of trays with product must be puncture-proof, extensible and impervious to gases like oxygen. Hundreds of different films are used in the packaging industry. These can be broadly categorised into two groups:

— basic films consisting of a single layer of film

— laminate consisting of two or more basic films glued together or bonded together by heat or by adhesives

Plastics such as polyethylene film or copolymer of ethylene and vinyl acetate are very often used for packing of frozen products. Polyethylene packs can be produced manually using pre-made bags. An impulse or bar sealer is used to seal the bags which are hand-filled. In order to improve the barrier properties of packages laminates are used, for example polyester/polythene.

Products which are particularly sensitive to oxygen are vacuum-packed. During the sealing operation, air is removed from the package. A laminate nylon/polythene is commonly used as packaging material. This type of packaging is used, for instance, for smoked trout which are arranged on a board with, for example, a coated texture. Numerous machines exist for vacuum-packing with single, double or continuous chambers. Vacuum-sealing machines can additionally be equipped with a modified atmosphere packing system (MAP).

Immediately on removing the air from the package a mixture of gases is pumped in. Usually this mixture consists of 30% nitrogen, 40% carbon dioxide, and 30% oxygen. In the case of fat fish the oxygen is replaced by nitrogen. This method is increasingly used for packing fresh fish. The MAP products have to be stored at the temperatures lower than 3°C because of C. botulinum hazard. MAP packages consist of two kinds of foil. The bottom film is foil-rigid or semi-rigid. This foil is formed by, for example, extrusion and the resultant tray is moved to the packing section. Because of product drip it is placed on an absorbing board.

The top web is drawn over the filled trays and sealed round the edges. The pack may be evacuated or gas-flushed before sealing. Vacuum-skin packaging is becoming more common for packing smoked fish. In this process the wrapper is heated and wrapped over the product, the film moulding completely to the product shape and sealing the product completely, forming an extra skin.

Labelling

Lack of detailed standards and existence of only limited regulations concerning wholesomeness and sanitary conditions for production and trade of food products characterise the market economy. Here, the problem of labelling is of a particular importance. Regulations in this regard are very detailed and are aimed at protecting the health of the consumer and providing the best information. These requirements enable the consumer to decide which products to buy. A label placed on the product should inform the consumer about the raw material used, method of preparation and form of consumption, shelf life, etc.

Establishments and Production Demands

Plant Location

Before deciding plant location different factors should be analysed. The most important is the plot which should be of adequate size for both present needs and future development. The plant should be close to public transport such as rail or road. Access to electricity, water, and steam is essential. Waste disposal should be considered when planning the plant location. The owner should coordinate all the works with local competent authorities in order to avoid problems in the future.

The choice of plant location should also take into account the neighbouring surroundings: for example, location near to a waste dump could lead to microbiological contamination caused by birds. A well designed building should comprise sufficient space for work to be conducted out under adequate hygienic conditions, an area for machinery, equipment and storage, separation of operations that might contaminate food, adequate natural or artificial lighting, ventilation, protection against pests. There are many technical regulations concerning construction of buildings and processing halls; *eg*., outside walls, windows and doors should be constructed such that they are water-, insect- and rodent-proof.

The inside walls of the building should be painted white or other light colour and their surface should be smooth, fall-safe, corrosion-proof and easy to clean. Floors should be resistant to spillage of products, water and disinfectants. They should be slip-proof and maintain their colour. Experience shows that selection and preparation of the floor is one of the most difficult tasks facing the designer.

The main problem, however, lies in appropriate general layout and arrangements of rooms which must minimise the risk of contamination of the final product. The majority of pathogens and spoilage micro-organisms derive from the raw material. In order to avoid cross-contamination the raw material should be placed in separate cold stores. The best solution seems to be separation by walling-off the unclean area from the clean area. The unclean area is where the raw material is delivered, sorted and possibly processed, *eg*., gutted.

Clean areas are places of production where any contaminants added to the product could be transmitted to the product, *ie*., there is no subsequent processing step that will reduce or destroy the contaminating microbes. Thus separation of these areas has to be complete and there should no movement of people or equipment from unclean to clean area. Proper layout and designs should ensure an uninterrupted and "straight line" process flow, and should meet other requirements listed below:

— all functions should avoid zigzagging and backtracking

— visitors should move from unclean to clean areas

— conditioned (chilled) air and drainage should flow from clean to unclean areas

— the flow of discarded outer packing material should not cross the flow of either unwrapped ingredients or finished product

— there should be sufficient space for plant operations including processing, cleaning and maintenance; space is also required for movement of materials and pedestrians

— operations are separated as necessary. There are clear advantages in minimising the number of interior walls since this simplifies the movement of materials and employees, simplified supervision, and reduces the area of wall that needs cleaning and maintenance

The proper design and arrangement of the processing plant greatly influence food production hygiene. Machinery safety and hygiene contains the following most important requirements:

— Machinery containing materials intended to come in contact with food must be designed and constructed so that these materials can be cleaned each time they are used.

— All surfaces and joints must be smooth, with no ridges or crevices that could harbour organic materials.

— Assembly must be designed so as to minimise projections, edges and recesses; they should be constructed by welding or continuous bonding, with screws, screwheads and rivets used only where technically unavoidable.

— Contact surfaces must be easy to clean and disinfect, and be built with easily dismantled parts; inside surfaces must be curved so as to allow thorough cleaning.

— Liquid derived from foods, and cleaning, disinfecting and rinsing fluids should be easy to discharge from machinery.

— Machinery must be designed and constructed to prevent liquids or living creatures - primarily insects - from entering and accumulating in areas that cannot be cleaned.

— Machinery must be designed and constructed to avoid ancillary substances, such as lubricants, coming into contact with food.

Personal Hygiene

Personal hygiene is a most important element of health quality assurance in a fish processing plant. The essential requirements for personnel working in production area and stores are described below:

1. Protective clothing, footwear and headgear issued by the company must be worn and must be changed regularly. When considered appropriate by management, a fine hairnet must be worn in addition to the protective headgear provided. Hair clips and grips should not be worn. Visitors and contractors must comply with this regulation.
2. Protective clothing must not be worn off the site and must be kept in good condition. If it is in poor condition the supervisor should be informed immediately.
3. Beards must be kept short and trimmed, and a protective cover worn when considered appropriate by management.
4. Nail varnish, false nails and make up must not be worn in production areas.
5. False eyelashes, wrist watches and jewellery (except wedding rings or the national equivalent, and sleeper earrings) must not be worn.
6. Hands must be washed regularly and kept clean at all times.

7. Personal items must not be taken into production areas unless carried in inside overall pockets (handbags, shopping bags must be left in the locker provided).
8. Food and drink must not be taken into or consumed in areas other than the tea bars and the staff restaurant.
9. Sweets and chewing gum must not be consumed in production areas.
10. Smoking or taking snuff is forbidden in food production, warehouse and distribution areas where 'No Smoking' notices are displayed.
11. Spitting is forbidden in all areas of the site.
12. Superficial injuries (cuts, grazes, boils, sores and skin infections) must be reported to the medical unit or nurse via the supervisor and clearance obtained before entering production areas.
13. Dressings must be waterproof and contain a metal strip as approved by the medical unit.
14. Infectious diseases (including stomach disorders, diarrhoea, skin conditions and discharge from eyes, nose or ears) must be reported to the medical unit or nurse via the supervisor. This also applies to staff returning from travel abroad where there could be a risk of infection.
15. All staff must report to medical unit on return from both certified and uncertified sickness.

Cleaning in Processing Plant

Water quality

As a general rule, water used for all purposes in food production must meet drinking water standards. Disinfectant residues should be monitored where possible and the bacteriological quality periodically checked. Turbidity, colour, taste and odour are also easily monitored parameters. If there are local problems with chemical constituents (fluoride, iron) or contaminants from industry or agriculture (*eg*., nitrate, pesticides, mining wastes) these should (hopefully) be monitored and dealt with by the water suppliers.

Very often water must undergo treatment disinfection prior to use. The following chemicals are used as disinfectants: chlorine, chloramine, ozone or UV irradiation. Chlorination is the cheapest form of treatment and monitoring of chlorine is relatively easy. According to WHO the concentration of chlorine in water should be in the range 0.2-0.5 mg/l. For sanitation purposes it may reach 200 mg/l, but in order to avoid corrosion lower concentrations are advised (50-100 mg/l).

Cleaning and disinfection are the most frequent operations in modern food processing. Carelessness may cause considerable economic loss, and loss of reputation on the market. The hygienic standards respected in processing plants depend on kinds of production. For example, in the cannery they will be more strict than in plants where fish is only gutted and stored in ice and its shelf life is rather short. Regarding all other technological operations and processes, cleaning and disinfection procedures must follow detailed instructions and responsible personnel be assigned. Various steps should be included in a complete cycle of cleaning and disinfection

1. Remove food products, clear area from bins, containers, etc.
2. Dismantle equipment to expose surfaces to be cleaned. Remove small equipment, parts and fittings to be cleaned in a specified area. Cover sensitive installations to protect them against water, etc.
3. Clear the area, machines and equipment of food residues by flushing with water (cold or hot) and by using brushes, brooms, etc.
4. Apply the cleaning agent and use mechanical energy (*eg*., pressure and brushes) as required.
5. Rinse thoroughly with water to completely remove the cleaning agent after the appropriate contact time (residues may completely inhibit the effect of disinfection).
6. Control of cleaning.
7. Sterilisation by chemical disinfection or heat.
8. Rinse off the sterilant with water after the appropriate contact time. This final rinse is not needed for sterilants, *eg*., H_2O_2 based formulations which decompose rapidly.
9. After final rinsing, equipment is reassembled and allowed to dry.
10. Control of cleaning and disinfection.
11. In some cases it will be good practice to re-disinfect (*eg*., with hot water or low levels of chlorine) just before production recommences.

Only agents and disinfectants permitted by adequate regulations, can be used for cleaning and disinfection operations. During their use precautionary measures must be observed and this requires proper training of personnel.

Quality Aspects

Public health aspects

The term quality has many different implications, *eg*., product excellence, value, nutrition, safety for consumer, etc. In a free market economy the producers are responsible for food quality and they are controlled by the competent authorities according to approved procedures. Certain countries or groups of countries, *eg*., European Union, formulate regulations specifying requirements concerning health quality, wholesomeness of raw materials and food/fish products and concerning permissible limits for chemical contaminants (heavy metals, PCBs, etc.) or biological infestants (parasites, microbes, etc.).

Other regulations concern quality of water provided for food processing. These regulations are of rather general character but there are others which concern health conditions for processing and placing of products on the market. Due to an almost complete lack of detailed standards for individual products, the regulations on labelling are of great importance, especially if the "fair trade" principle and consumer interest are to be taken into account. All these groups of obligatory regulations should ensure production of food which is safe for the consumer.

Additionally the monitoring of raw materials is a complementary part of activities carried out according to requirements contained in regulations. It provides the competent authorities, responsible for supervision of production, with information about potential hazard. Producers are responsible for food quality.

Veterinary inspection should cover the visual assessment of aquaculture fish wholesomeness, taking of fish samples and immediately sending them to the competent laboratory. Each farm must record all necessary data pertaining to the wholesomeness of fish including the official certificate of laboratory analysis. It is pointed out that only certificates issued by official control authorities are valid and placing of fish in the fish farm or for sale has to be formally documented.

Fish processing plants which, apart from farming carry out processing must obtain the approval of veterinary authorities to export their products. Acquiring a registration number is a formal approval to export fish. This registration number enables identification of the fish product, and this number must be shown on the label of each package and on the relevant documents.

Quality control

Control is traditionally limited to control of the final product. Practice has proved that this is not sufficient and that quality control should be carried out during all stages of production, starting from a contract on supply of raw material, through all the phases of processing, to storage and distribution of final products. Such an approach is not quality control but constitutes quality assurance, which covers the entire production chain. Below, principles relating to quality control are presented with regard to the main individual operations and procedures in fish processing.

Drawing up a contract for raw material supply: The contract for supply of raw material should cover all specific requirements, for example: size of fish, closed seasons, level of chemical contaminants in fish and in the water from which the fish comes, and chemical measurements should be made by an institution which deals with monitoring of environment. Sometimes, especially in the case of export, the buyer/customer may have additional demands, *eg*., an indication of the level of chemical contaminants other than standard ones.

The buyer should ensure that he will receive the health certificate for his raw material and that the certificate was issued by the official control authorities. He should also obtain confirmation that the fish was stored properly prior to sale (for example, that fish was iced with a proper amount of ice and that the quality of the ice was satisfactory; that it was stored in cool store rooms and that it was transported by appropriate means). The contract may specify that some of these demands be passed to the receiver.

Receiving and storage of raw material: This control step determines the quality of the final product and should be carried out extremely carefully. In general it consists of three elements:

- temperature control of fish during transportation (temperature record)
- temperature control of fish and control of icing
- quality control of purchased fish

The temperature of the fish hold in the transport system is usually registered automatically or periodically by a driver. This temperature record is part of the documentation on fish shipment. Measurements of real temperature of fish tissue and control of icing are made on random samples. Apart from these elements the cleanliness of the means of transport and the containers, and the labelling, are checked. The number of samples/packages with fish to be further assessed depends on lot size, and it should be specified clearly in compulsory procedures or codes of good manufacturing practice, perhaps in the standards or contract specifications.

The temperature of purchased fish should be close to ice melting temperature and not higher than 4°C. The samples of fish taken for temperature measurement are at the same time the samples examined for quality control of raw material. Usually in the case of medium size batches eight packages are taken and in each package three temperature checks are made. Detailed quality assessment is made according to requirements laid down in procedures, codes or standards if the latter exist. Such an assessment is carried out on an average sample from a set of randomly selected packages.

The sensory analysis of raw material is a main part of control, and it allows full characteristics of the fish investigated to be obtained. This analysis includes appearance of skin, eyes, gills and fish as a whole, colour of fish tissue; damage to fish, springiness of meat tissue, flavour of individual organs; flavour, taste and texture of meat tissue after cooking. Occurrence of inadmissible features like for example sour smell of gills, strange/ unfamiliar smell of meat or fish as a whole causes that raw material is disqualified and excluded as a material intended for processing.

In the case of live fish their appearance and movement in the water in a container are assessed. The kind and the degree of infestation with parasites determines further procedure. If the presence of parasites which are harmful for humans is detected, fish cannot be sold as fresh. Generally when fish is qualified as conforming with requirements and cooled properly it is placed in cold stores or transported direct to the processing line. Ice is added to fish cooled insufficiently and this is placed in cold store. The temperature inside the cold store should be close to 0°C and should be continuously recorded. If temperature cannot be registered automatically, measurements should be taken not less frequently than every two hours.

Quality control during the production process: The quality control programme during the production process depends on the profile of production carried out in the processing plant. Each processing plant must draw up a flow chart of the entire process starting from the raw material through every individual operation and process to the final product and with all quality control points indicated. Criteria for selection of control points depend on potential hazards which, in the case of lack of proper handling, can cause a risk for both the food and the consumer. For example, control of temperature during individual operations, their duration, concentration of food additives, etc., are typical and critical parameters measured at control points.

Technological supervision is responsible for use of adequate processing parameters. Quality control personnel are responsible for monitoring these parameters and in the case of deviation they should undertake proper corrective action. The final step in production control is the quality control of the final product according to technical requirements and specifications included in the contract or standards if the latter are compulsory. Such assessment is carried out according to approved procedures with special regard to health quality requirements pertaining in a given country. This type of control will disappear in the future because an introduction of quality assurance systems, as a continuous control throughout the entire processing procedure, will eliminate this traditional form of control.

Quality control personnel are also responsible for supervision of assurance of cleanliness and disinfection of production lines and processing rooms. Maintenance of cleanliness and disinfection should be carried out in accordance with a programme approved by the local veterinary service. The quality control staff assures adherence to this programme which especially concerns:

— types of detergents/disinfectants and concentrations used;

— compliance with procedures of cleaning/washing and disinfection;

— arrangement of periodic microbiological measurements on the surface of equipment and processing machines;

— control of personal hygiene of staff including working clothes and sanitary fittings in the plant.

Storage and distribution of freshwater fish products: The fish products directed for the storage or for the purchaser are random checked by the internal quality control staff. This control, inter alia, concerns:

— proper packaging materials and labelling (according to official requirements);

— duration and temperature of storage;

— proper conditions of storage, for example adequate ice, temperature etc;

— choice of means of transportation and hygienic conditions (cleanliness, temperature record, etc.);

— proper loading (*eg*., arrangement of load in vehicles).

Polyphosphates in Fish Process

Polyphosphates are legally permitted additives that are widely used to aid processing or to improve eating quality of many foods, particularly meat and fish products. Phosphates are also used in making baking powder and cola drinks, and great quantities are used in fertilisers and detergents. Phosphates are present normally in all living things and are an essential component of our diet. A phosphate is a salt of phosphoric acid; when a number of simple phosphate units are linked to form a more complex structure, this is known as a polyphosphate. The phosphates used in foods may be simple phosphates, pyrophosphates containing two phosphate units, tripolyphosphates containing three units, or polyphosphates containing more than three phosphate units.

Benefits of Polyphosphates

Many benefits have been claimed to result from the use of polyphosphates. The main value of polyphosphates lies in improving the retention of water by the protein in fish. The manner in which they do this is not clearly understood, but their effect is mainly on the surface of the fish. Other substances such as common salt give similar results but with undesirable flavour effects. Polyphosphate treatment of fish before freezing often reduces the amount of thaw drip, that is the liquid released when frozen fish is thawed.

Good quality fish, properly frozen and cold stored, normally develops little thaw drip; therefore application of polyphosphate to such material is generally only of slight value. Poor quality fish when frozen and thawed may drip much more, and treatment will reduce the loss to some extent, but this is not sufficient reason for using polyphosphates; poor quality fish should not be frozen, since the product will be poor irrespective of treatment. Good quality fish that is frozen and then stored at too high a temperature may have a high thaw drip loss, and again polyphosphate treatment before freezing can reduce the loss, but this does nothing to prevent the corresponding deterioration in flavour and texture; there is no substitute for good cold storage.

Many retail products, especially breaded ones, are cooked from the frozen state without prior thawing; thaw drip from such products is generally of no great significance and the value of phosphating is doubtful. Chilled fillets lose water slowly during processing and distribution. The loss is usually small, but it may be increased by high temperature, excessive pressure on the product, and by delay in distribution. Polyphosphate treatment of the fillets will reduce drip loss, but will have no appreciable action on the effects of spoilage on odour and flavour; treatment does not reduce the need for rapid handling or the need for keeping the fish chilled. There are two instances where polyphosphate treatment of chilled fillets or portions may have a specific advantage.

First, chilled fillets of good quality from thawed sea frozen fish often lose a high amount of drip during distribution; polyphosphating after filleting can substantially reduce this loss, and can also dramatically improve the appearance of the fillets by changing the dull, often brownish, surface that is typical of sea frozen material to a glossy bluish white. The dull surface of poor quality fillets from stale fish can also be given a gloss by treatment with polyphosphate, but it must be emphasised that the improvement is only in appearance; the poor quality remains unchanged.

Secondly, polyphosphate treatment can improve the appearance of prepacked chilled fillets by preventing the accumulation of unsightly drip within the pack. Many retail products such as fish fingers and portions are cut from frozen blocks of bonefree flesh known as laminated blocks. The preparation of these blocks from boneless, skinless fillets is simplified by polyphosphate treatment. The polyphosphate swells some surface protein to give a lubricating layer which enables the fillets to be packed easily into moulds, without leaving voids which could affect the shape and weight of products cut from the blocks.

However, polyphosphates are not an essential ingredient of laminated blocks; quantities of blocks have been manufactured without addition of polyphosphates. Blocks are also made from minced fish flesh, or a proportion of mince may be incorporated in fillet blocks. Polyphosphate mixed with the mince acts as a binding agent by swelling protein to give the product an improved texture. A similar binding effect occurs when the treatment is applied to fillets that are particularly liable to break up on thawing, for example hake fillets.

The use of polyphosphates in laminated blocks cannot be justified on the grounds that drip is reduced; not only is thaw drip not a problem with products cut from laminated blocks, but the action of cutting a block greatly reduced the antidrip effect of the polyphosphate on the surface of the fish. The first and universal effect of all polyphosphate treatment is to increase the weight of the fish by retaining water. The weight gain is of no technological benefit, and represents to the producer a gain in weight of product sold. This weight gain, which amounts to selling added water, clearly must not be the aim of polyphosphate treatment; polyphosphates should be added to fish only for technically justifiable purposes. Polyphosphate treatment makes fillets slippery and difficult to handle; although slipperiness is an advantage when making laminated blocks, it may be a nuisance under some circumstances.

Polyphosphates cannot significantly improve the eating quality of fish, although claims in this respect are often made. Excessive treatment of small products such as shellfish or thin fillets can even result in undesirable flavour changes and sloppy texture. Polyphosphates have no preservative action on chilled fish, nor do they give any protection against cold store deterioration. Most processors now use proprietary mixtures containing appropriate polyphosphates. These mixtures may contain pyrophosphate, tripolyphosphate or compounds with more than three phosphate units; these last have the advantage of being highly soluble in water, so that the supplier can offer concentrated solutions that are readily diluted to the required strength by the processor.

Application

Proprietary solutions should be prepared according to the maker's instructions. If sodium tripolyphosphate is used as the active ingredient, a satisfactory mixture can be made by dissolving 5 kg of food grade sodium tripolyphosphate and 5 kg of glassy sodium phosphate in 90 litres of water. The solution should be chilled to avoid warming the fish and to increase the life of the solution. The simplest way of applying polyphosphate is to immerse a batch of fillets or pieces of fish contained in an open mesh basket, agitating them gently so that all surfaces come into contact with the solution. Alternatively the fish can be conveyed mechanically through a long shallow tank of solution; suitable machines are available that can be incorporated in a continuous processing line. The solution should be renewed daily, or more often if badly discoloured or contaminated. A third possibility is to tumble the fillets gently in a slowly rotating drum containing a small amount of solution, say 5 per cent of the weight of the fish; most of the liquid is taken up by the fish in a few minutes.

Minces can be treated by adding either solution or dry powder at some stage in the mincing or mixing procedure. Polyphosphate treatment should be just enough to produce the desired technological effect and no more. The solution prepared as described above contains 5 per cent of the active ingredient, sodium tripolyphosphate; weaker solutions have less ability to reduce drip although they have some value, for example in improving the appearance of fillets of sea frozen fish or in helping to make good fillet blocks. Immersion time in a batch or continuous process should be no longer than necessary; usually a minute or so is enough. Longer immersion time leads to unjustified increase in weight and the risk of flavour or texture deterioration. Since polyphosphate acts on the surface of fish, fillets should not be cut or trimmed after treatment.

Most polyphosphates added to food are broken down to single phosphate units in the stomach when the food is eaten; indeed, many are converted to single units in the food before it is eaten, for example in chill storage or during cooking. Thus it is clear that most phosphates added to food are nutritionally equivalent to the phosphates naturally present in food, and are likely to present little hazard to health. Nevertheless it is possible for essential food components to be harmful if taken in excess, and international medical authorities have recommended that the daily consumption of all phosphates should not exceed a certain level. It is believed that the amounts presently being eaten by the typical consumer are well below this limit. Some countries do have limits on the amount that can be added and exporters must bear this in mind.

References

Bekker-Nielsen T. (2005). *Ancient fishing and fish processing in the Black Sea region* Volume 2 of Black Sea studies, Aarhus University Press.

Bremner HA. (2003). *Safety and Quality Issues in Fish Processing* Woodhead Publishing Limited.

Luten JB, Jacobsen C and Bekaert K. (2006). *Seafood research from fish to dish: quality, safety and processing of wild and farmed fish* Wageningen Academic Publishers.

Stewart H. (1982). *Indian Fishing: Early Methods on the Northwest Coast* University of Washington Press.

United Nations Development Fund for Women. (1993). *Fish processing* Food Technology Source Book Series (UNIFEM) Series.

Bibliography

Alasalvar C (2010) *Seafood Quality, Safety and Health Applications* John Wiley and Sons, Page 203.

Alzamora SM, Tapia MS and López-Malo A (2000) *Minimally processed fruits and vegetables: fundamental aspects and applications* Springer.

Bekker-Nielsen T. (2005). *Ancient fishing and fish processing in the Black Sea region* Volume 2 of Black Sea studies, Aarhus University Press.

Bremner HA. (2003). *Safety and Quality Issues in Fish Processing* Woodhead Publishing Limited.

Campbell, Bernard Grant. (1998). *Human Evolution: An Introduction to Man's Adaptations*. Aldine Transaction:

Carcinogens and Anticarcinogens in the Human Diet. National Academy Press. (1996).

Carpenter, Ruth Ann; Finley, Carrie E. (2005). *Healthy Eating Every Day*. Human Kinetics.

Chan, H.T. Jr. (1984). *Handbook of Tropical Foods*. New York/Basel: Marcel Dekker.

Christine Dell'Amore in Chicago. (2009). "Cooking Gave Humans Edge Over Apes?". National Geographic News.

Colin, D. (1992). Recent trends in fruit and vegetable processing. In *Food Science and Technology Today,* 7, (2)), pp. 111 - 116.

Corpet DE, Yin, Zhang XM, *et al.* (1990). "Colonic protein fermentation and promotion of colon carcinogenesis by thermolyzed casein". *Nutr Cancer* 23 (3): 271–81.

Curley, S., and Mark. (1990). *The Natural Guide to Good Health*, Lafayette, Louisiana, Supreme Publishing.

Davidson, Alan. (2006). *The Oxford Companion to Food.* 2nd ed. UK: Oxford University Press.

Desrosier, N.W. and Desrosier, J.N. (1977). *Technology of Food Preservation*, 4th Edition, AVI Publishing Co. Westport, Conn.

FAO: Preservation techniques Fisheries and aquaculture department, Rome. Updated 27 May 2005. Retrieved 14 March 2011.

Food and Agriculture Organization of the United Nations. *The State of Food Insecurity in the World 2005*. . Retrieved fromhttp://www.fao.org/docrep/008/a0200e/a0200e00.htm on 2006-09-29.

Galdston, I. (1960). *Human Nutrition Historic and Scientific.* New York: International Universities Press.

Gorman, RM.(2008). "Cooking up bigger brains". *Scientific American* 298 (1): 102, 104–5.doi:10.1038/scientificamerican0108-102.

Jango-Cohen, Judith. (2005). *The History of Food.* Twenty-First Century Books.

Jin, Y.; Wang, M.; Rosen, R. T.; Ho, C. T. (1999). "Thermal Degradation of Sulforaphane in Aqueous Solution". *Journal of Agricultural and Food Chemistry* 47 (8): 3121–3123.

Katz, Solomon. (2003). *The Encyclopedia of Food and Culture*, Scribner.

Kripke, Gawain. (2005). *Food aid or hidden dumping?*. Oxfam International, March.

Lee S (2004) "Microbial Safety of Pickled Fruits and Vegetables and Hurdle Technology" *Internet Journal of Food Safety*, 4: 21-32.

Leistner L (1995) "Principles and applications of hurdle technology" In Gould GW (Ed.) *New Methods of Food Preservation*, Springer, pp. 1-21.

Luten JB, Jacobsen C and Bekaert K. (2006). *Seafood research from fish to dish: quality, safety and processing of wild and farmed fish* Wageningen Academic Publishers.

Mahan, L.K. and Escott-Stump, S. eds. (2000). *Krause's Food, Nutrition, and Diet Therapy* (10th ed.). Philadelphia: W.B. Saunders Harcourt Brace.

Marion Nestle: *Food Politics: How the Food Industry Influences Nutrition and Healt*h, University Presses of California.

McGee, Harold. (2004).*On Food and Cooking: The Science and Lore of the Kitchen.* New York: Simon and Schuster.

Mead, Margaret. (1997). *The Changing Significance of Food.* In Carole Counihan and Penny Van Esterik (Ed.), Food and Culture: A Reader. UK: Routledge.

Regmi, Anita (2001).*Changing Structure of Global Food Consumption and Trade.* Market and Trade Economics Division, Economic Research Service, USDA, May 30.

Stafford, A.E. (1984). "Mango". In *Handbook of Tropical Foods*, pp. 399-432. H.T. Chan, Jr. Editor. New York/Basel: Marcel Dekker

Stewart H. (1982). *Indian Fishing: Early Methods on the Northwest Coast* University of Washington Press.

Thiollet, J.-P. (2001). *Vitamines & minéraux*. Paris: Anagramme.

Thompson, A.K. (1989). *Recent advances in post-harvest technology of fresh fruits and vegetables.* Cranfield Institute of Technology, Silsoe College, U.K.

United Nations Development Fund for Women. (1993). *Fish processing* Food Technology Source Book Series (UNIFEM) Series.

Yousef AE and Carolyn Carlstrom C (2003) *Food microbiology: a laboratory manual* Wiley.